Game-Playing for Active
Ageing and Healthy Lifestyles

RIVER PUBLISHERS SERIES IN INFORMATION SCIENCE AND TECHNOLOGY

Series Editors:

K.C. CHEN
National Taiwan University, Taipei, Taiwan
and
University of South Florida, USA

SANDEEP SHUKLA
Virginia Tech, USA
and
Indian Institute of Technology Kanpur, India

The "River Publishers Series in Information Science and Technology" covers research which ushers the 21st Century into an Internet and multimedia era. Multimedia means the theory and application of filtering, coding, estimating, analyzing, detecting and recognizing, synthesizing, classifying, recording, and reproducing signals by digital and/or analog devices or techniques, while the scope of "signal" includes audio, video, speech, image, musical, multimedia, data/content, geophysical, sonar/radar, bio/medical, sensation, etc. Networking suggests transportation of such multimedia contents among nodes in communication and/or computer networks, to facilitate the ultimate Internet.

Theory, technologies, protocols and standards, applications/services, practice and implementation of wired/wireless networking are all within the scope of this series. Based on network and communication science, we further extend the scope for 21st Century life through the knowledge in robotics, machine learning, embedded systems, cognitive science, pattern recognition, quantum/biological/molecular computation and information processing, biology, ecology, social science and economics, user behaviors and interface, and applications to health and society advance.

Books published in the series include research monographs, edited volumes, handbooks and textbooks. The books provide professionals, researchers, educators, and advanced students in the field with an invaluable insight into the latest research and developments.

Topics covered in the series include, but are by no means restricted to the following:

- Communication/Computer Networking Technologies and Applications
- Queuing Theory
- Optimization
- Operation Research
- Stochastic Processes
- Information Theory
- Multimedia/Speech/Video Processing
- Computation and Information Processing
- Machine Intelligence
- Cognitive Science and Brain Science
- Embedded Systems
- Computer Architectures
- Reconfigurable Computing
- Cyber Security

For a list of other books in this series, visit www.riverpublishers.com

Game-Playing for Active Ageing and Healthy Lifestyles

Dr. Ana Isabel Veloso

University of Aveiro-DigiMedia, Aveiro, Portugal

Dr. Liliana Vale Costa

University of Aveiro-DigiMedia, Aveiro, Portugal

River Publishers

Routledge
Taylor & Francis Group

LONDON AND NEW YORK

Published 2021 by River Publishers
River Publishers
Alsbjergvej 10, 9260 Gistrup, Denmark
www.riverpublishers.com

Distributed exclusively by Routledge
4 Park Square, Milton Park, Abingdon, Oxon OX14 4RN
605 Third Avenue, New York, NY 10158

First published in paperback 2024

Game-Playing for Active Ageing and Healthy Lifestyles / by Ana Isabel Veloso, Liliana Vale Costa.

Routledge is an imprint of the Taylor & Francis Group, an informa business

Publisher's Note
The publisher has gone to great lengths to ensure the quality of this reprint but points out that some imperfections in the original copies may be apparent.

While every effort is made to provide dependable information, the publisher, authors, and editors cannot be held responsible for any errors or omissions.

ISBN: 978-87-7022-437-6 (hbk)
ISBN: 978-87-7004-290-1 (pbk)
ISBN: 978-1-003-33824-6 (ebk)

DOI: 10.1201/9781003338246

"Nobody grows old merely by a number of years. We grow old by deserting our ideals."

–Samuel Ullman

Acknowledgement

The editors want to thank all the participants and Universities of the Third Age that collaborated with us in this project. A special thanks to our students who contributed with their prototypes and research to advance the Project SEDUCE 2.0–Use of Communication and Information in the miOne online community by senior citizens. This work was supported by FCT (Fundação para a Ciência e Tecnologia), COMPETE 2020, Portugal 2020 and EU through the European Regional Development Fund-the project SEDUCE 2.0 nr. POCI-01-0145-FEDER-031696.

–Ana Isabel Veloso
–Liliana Vale Costa

Contents

Foreword

Our lives are characterized by continuous and sometimes disruptive transformations. Many aspects of our lives change during our lifespan, in particular when it comes to aging. While it is hard to specify exactly when and how we need to gain knowledge to adapt to novel situations and/or technologies, playing and serious gaming, in the right context, helps to keep up our performance and strengthen memory while accumulating heterogeneity of experiences.

The authors of this book introduce constructive ways in which game-based tools can be designed and used to improve our cognitive and social lives when aiming to age actively. Taking a user-centered design perspective, the authors provide a synthesis of the challenges accompanying development across different disciplines. Their illustrative style helps in understanding the specific purpose of projects and endeavors to target market opportunities, sector-relevant particularities, and multiple stakeholders from different areas and communities in the context of gamification and respective tool development.

Readers with different backgrounds should easily grasp the contribution of effective game designs to overall wellbeing and specific aspects of older adults' health. The research-based evidence that playing games can help improve various aspects of health of older adults should encourage readers to engage in gamification, game application, or tool design processes, regardless of whether they act as caretakers, users, or developers. Design also is affected by accessibility issues that need to be taken into account for improving wellbeing and quality of life for older adults.

The presented methods and instruments relate well to gerontechnology, acknowledging the increasing propagation of digital systems into support tools and environments for aging adults. Of particular importance is the finding that games have the potential to be recognized as non-pharmacological interventions, facilitating rehabilitation processes and brain training effectively. However, customization seems to be crucial. A high degree of self-interest and engagement of users seems to be required when exploring and finally utilizing the full potential of gerontechnology tools and instruments. The more concisely the requirements are articulated, the more effective the selection of features and constellations will be during aging.

I recommend this book to everyone interested in the current state of affairs of game-playing for active aging support. The addressed issues provide orientation as well as insight into the most prominent topics and application areas at the interface of serious gaming and active aging support.

Professor Christian Stary
Head of the Department of Business Information Systems,
leading the Communications Engineering Team, at
Johannes Kepler University Linz, Austria

List of Figures

List of Tables

List of Abbreviations

ICT	Information and Communication Technologies
KPI	Key Performance Indicators
ROI	Return on investment
PvE	Person-versus-Environment
PvP	Person-versus-Person
AT	Assistive Technology
OECD	Organization for Economic Co-operation and Development
TRAP	Tremors, Rigidity, Akinesia, Postural instability
ICF	International Classification of Functioning
UI	User Interfaces
DDR	Dance Dance Revolution
FPS	First Person Shooters
MMORPGs	Massive Multiplayer Online Role-Playing Games
MDA	Mechanics-Dynamics-Aesthetics
DMC	Dynamics-Mechanics-Components
POIs	Points of Interest

1

Introduction

The Information and Communication Society has challenged the way we make daily-life decisions, interact with the environment and participate in economic, cultural, and social affairs. If, on the one hand, the use of digital platforms in domestic spaces and an increasingly service-driven market have facilitated the access to the information and task performance in time, on the other hand, intergenerational gaps can be generated regarding age, level of participation in the labor workforce, health status, and learning pace.

Alongside this reality, the global demographic crisis has heightened the need for developing and studying the design of digitally mediated tools to foster active ageing.

In addition, over the past few years, digital games have evolved from a mere entertainment industry to play an important role in transforming well-being economy, health-related behavioral changes, and even social care delivery. Such expressions and terms as "Games for a change," "Health games," "Games for well-being," "Gamifying behaviors," "Games for serious purposes," "Games for good," "Newsgames," and "Game-based learning" have been echoing game research very lately mostly grounded on the advances in cognitive sciences and psychology.

If games have been often regarded as an addictive violent activity throughout time, nowadays, this perspective is falling behind the potential benefits of games, namely in terms of cognition, rehab, and motivation to behavioral change.

As populations age, brain-training and rehabilitation games (e.g., Brain Age: Train Your Brain in Minutes a Day! Dr. Kawashima's Body and Brain Exercises) have formed the basis of a growing market. The game industry has also adopted the strategy of widening the target audience to all ages with the advent of family games and the use of such game consoles as Wii.

1

Although there is a significant potential for commercializing these game-based approaches that help with the ageing process given the larger target market, there is a need to understand the motivations and behaviors of the older adult consumers, the design process applied to behavioral theories, its inherited challenges to the Human–Computer Interaction field, and the reciprocal affordances that can co-exist between the end users and own environment (i.e., lifestyles and independent living, mobility, cities, health, and social networks).

In the video game industry, older adults are likely to become the next generation of video gamers, valuing cognitive player-versus-environment (PvE) challenges. Digital play has also become more and more important as leisure for a growing number of older adults.

While the mainstream game industry has been challenged with the emergence of an older target group, there have been also advancements in gamification and the proliferation of smart devices. As a matter of fact, the introduction of digitally mediated interactions in mobility and travelling have heightened the need to reinvent the Senior Tourism sector to reinforce the connection of older adult travelers with a certain place and enhance their pre-, in-loco, and post-experience.

Games are also likely to have a therapeutic effect by being cognitively challenged and rewarded, but still the game industry is not fulfilling this target group motivations and accessibility needs.

In addition, most of the research in the field has focused on the physical and cognitive effects of video games in the aging process. Up to now, the use of other active ageing dimensions that go beyond the health domains (i.e., sense of security, and participation in society)[1] in games addressed to this target group remain unexplored.

This book gives an account of challenges and opportunities of the Game Market and Silver Economy, the considerations for designing Game-based tools for Active Ageing, the use of Gamification in Senior Tourism and wellness-oriented products and procedures to undertake when assessing Games for Active Ageing.

The book was inspired by our previous research carried out on Geron-technology. Much of our 8-year experience comes from the development of the following research projects: SEDUCE 2.0–Use of Communication and

---

[1]Please check the World Health Organization's Framework about Active Ageing-Health, Security and Participation in Society on Active Ageing: a policy framework https://extranet.who.int/agefriendlyworld/wp-content/uploads/2014/06/WHO-Active-Ageing-Framework.pdf (Access date: August 17, 2021)

Information in the miOne online community, SERIOUSGIGGLE–Games for triggering a healthy ageing and lifestyle, and SEDUCE–Senior Citizen Use of Computer-mediated Communication and Information in Web Ecologies. In these projects, we involved different end-users, adult learners at the Universities of the Third Age and older adults at retirement homes, in the design process of technology-mediated tools (i.e., games, online courses, online community, tourism, health, and communication services). During the field research, we observed that games had a therapeutic effect by being cognitively challenged and rewarded but, at the same time, the game industry was not fulfilling these end users' motivations, gameplay context, and accessibility needs.

Our goal for this book is, therefore, to discuss the way games can be designed and used for active ageing and healthy lifestyles, providing insight into research from a range of areas: Information and Communication Sciences, Gerontechnology, Marketing and Advertising, Game Studies, Tourism, Health, and Cognitive Psychology. It should be suitable for students of these areas who wish to envision future programs or products for the older adult population, policy-makers, researchers, game designers, caregivers, gerontology and health professionals, and practitioners.

The book is organized thematically around the main topics: The Game Market and Silver Economy; Designing Game-based tools for Active Ageing, Gamification, Senior Tourism and the Wellness Market; and Assessing Games for Active Ageing.

The chapter "The Game Market and Silver Economy" provides an overview and case study insights from the "Silver Economy," the older adult consumer and the strategies adopted by the advertising agencies. It ends with the challenges and opportunities of addressing the game products to this target group.

"Designing Game-based tools for Active Ageing" is essential to understand what is meant by ageing actively in the Information and Communication Society, the design process for good, well-being and quality of life, different applications of technology-led products and services to an ageing population and, more specifically, the use of games to deal with degenerative diseases, rehabilitation from stroke and exercise both mind and body. This chapter also covers accessibility in games.

In "Gamification, Senior Tourism and the Wellness Market," the concept of Gamification is introduced and its potential to the segment of Senior Tourism and Wellness Market is discussed.

Finally, the chapter "Assessing Games for Active Ageing" reviews a set of methods and instruments for data collection that can be used to assess the effectiveness of games for active ageing.

The information presented not only relies on case studies, but also on the authors' previous experience in co-designing digitally mediated products and opens up an avenue into a set of topics that Human–Computer Interaction designers and gerontechnology researchers and students need to take into account in the wellness market.

2

Games in the Silver Market

This chapter discusses the popularity of the silver market, that is 50+ targeted products or services. It discusses possible applications of gerontechnology within the graying or silver market. Then, it presents a brief overview of the older adult consumer and advertising portrayals. Finally, this chapter addresses the challenges and opportunities of the ageing game market.

2.1 Developing Products for the Silver Market

Over the past years, there has been much interest in addressing products and services to the ageing population. Never has been the "graying" or "silver market" so popular and promising given the demographic changes in society, that is the rapid growth in the ratio of older adults per youngsters.

The term "graying" or "silver market" refers to the segment and marketing-oriented strategies to people aged 50 and over (Kohlbacher & Herstatt, 2011). Japan is one of the top-lead silver markets, leveraging products and services that meet the users' needs, motivations, routines, and mobility (Reinmoeller, 2011).

A user-centered design (UCD) approach is usually followed in this market in order to ensure a positive experience. According to the International Organization for Standardization (ISO) (ISO 9241-210:2010, 2010), the user experience may be defined as:

> "A person's perceptions and responses resulting from the use and/or anticipated use of a product, system, or service." - ISO 9241-210:2010, 2010 (en), Ergonomics of Human-System Interaction Part210: Human-Centered Design for Interactive Systems.

As the world population ages and this ageing process involves a number of biological, psychological, and social changes (Baltes & Carstensen, 1999; Dziechciaz & Filip, 2014; Fisk, Czaja, Rogers, Charness, & Sharit, 2009), implications on digitally mediated design often arise-for example, increase the size of the interface elements, have contrast, enable changes in keyboard settings, and guarantee the visibility of close contacts (Hanson & Crayne, 2005; Hawthorn, 2000). However, ensuring age-neutrality is also fundamental, that is avoid age bias and stereotypes relative to age-ageism.

According to Stroud (2005), there are some rules to approach age-neutral marketing. In specific, these are:

- Avoid bias, stereotypes, and assumptions relative to consumers' age and rely solely on facts to communicate the message and run marketing campaigns.
- Regularly review the consumer strategy adopted to be age-neutral.
- Assume no influence of age in consumers' behaviors, reactions to brands, and experiences.

When addressing this market in the Information and Communication Technologies sector, businesses have been focused on the following themes: home robots (e.g., Pino, Boulay, Jouen & Rigaud, 2015; Smarr, Fausset & Rogers, 2011), health monitoring and self-care (e.g., Li, Ma, Chan & Man, 2019; Liu, Stroulia, Nikolaidis, Miguel-Cruz, & Rincon, 2016, VandeWeerd et al., 2020), and tourism (Nimrod, 2012; Pesonen, Komppula & Riihinen, 2015).

The development of Gerontechnologies has become key instruments in older adult's autonomy, self-care, and confidence in performing daily-life activities (e.g., Damuleviciene, Lesauskaite, Knasiene & Macijauskiene, 2010; Sale, 2018). Bouma, Fozard, Bouwhuis & Taipale (2007) use the term "gerontechnology" to refer to the technology, which aims to meet the challenges of increasing demands of accessing to digitized services and using digital platforms to support daily-life activities, proper of a digital but ageing society. Furthermore, the same authors (2007) draw our attention to the following domains that are contained in the umbrella term of gerontechnology:

- *Gerontology*: This field of studies delves into the human ageing process, namely the biological, psychological, and social aspects of ageing (Grabinski, 2007). Such themes as physiology, nutrition, psychology, and social psychology are covered.
- *Sociology*: In this field, themes include theories on individual and societal perspectives on ageing, demography changes in society, age

stratification, and social relationships (Marshall & Bengtson, 2011; Settersten & Angel, 2011).

- *Technology*: This field encompasses Chemistry-Biochemistry (e.g., Bouma, 1998), Architecture Building (e.g., Liddicoat & Newton, 2019), Mechatronics, Robotics, Ergonomics-Design (Czaja, Boot, Charness, & Rogers, 2019; Shishehgar, Kerr & Blake, 2018); and Business-Management (Lian & Yen, 2014; Moschis, 2003).

Although these domains are essential to gerontechnology, a cross-disciplinary approach is considered instead of a transdisciplinary one. This constitutes a major drawback, and a holistic approach would serve the cohesiveness and interrelatedness of the different domains.

Figure 2.1 illustrates the definition of the broad term Gerontechnology and all the three main subjects that are involved and interrelated: Anti-ageing, Assistive/Adaptive, and Enhanced/Ameliorated ageing technologies.

In our perspective, gerontechnology would be best described as the science that studies the convergence of different techniques and production of goods or techniques that can act upon the ageing process (i.e., sensation, cognitive processes, or operative skills). Furthermore, the different types of technologies that vary in terms of purpose (i.e., anti-ageing, assist/adapt, or enhance/ameliorate) rely on the exchange of different knowledge and insights to act in ageing. In specific, the different types of technologies are described below:

- *Anti-ageing technologies* are technologies that aim to delay and optimize the ageing process. Examples include regenerative medicine (implants, replace human organs, and molecular repair).
- *Assistive/Adaptive ageing technologies* refer to the technologies that aim to assist the older adults to perform daily-life activities. Examples include age-smart buildings, home, cities, workplaces, and transportation; companion robots, digitally-mediated rehabilitation programs, and ICT-based solutions for caregiving.
- *Enhanced/Ameliorated ageing technologies*, which intend to improve the older adults' skills and encourage active ageing.

It is worth noting, however, that these different types of technologies have been challenged by such aspects as: ethics, technology phobia and cost, accessibility, and user-centered design, the need to create literate and inclusive environments, technology determinism, and the challenges of ageism and its application in technologies (techno-ageism) and games (game-ageism) (Vale Costa, Veloso, Loos, 2019).

γέρων
Greek ageing

τέχνη
Greek technique

λόγια
Greek logos, discourse,
language, theory

Definition

science that studies the convergence of different techniques
and/or production of goods tha can act upon the ageing process (i.e. sensation, cognitive process or operative skills).

Domains

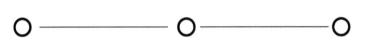

Anti-ageing

Technologies that aim to
delay or optimize the
ageing process.

e.g.: regenerative medi-
cine (implants, replace
human organs, molecular
repair)

Assistive/Adaptative

Technologies that aim to
assist the older adults to
perform activities in daily
living.

e.g.: age-smart buildings,
home, cities, workplace,
transportation, compan-
ion robots, rehab ICT.

Enhanced/Ammeliorated

Technologies that aim to
improve the older adults'
skills and encourage
active ageing.

e.g.: Cognitive and physical training technologies,
games, eHealth, Occupa-
tional therapy...

Figure 2.1   Domains of gerontechnology.

In the next section, older adult consumers and the way they are portrayed
in advertising will be discussed.

2.2 The Older Adult Consumer and Media Advertising

Although there has been a great interest in silver market given the worldwide
ageing population, the older adult continues to be an under-represented
segment in the advertising industry and media representations (Hettich,
Hattula & Bornemann, 2018).

The social-emotional selective theory ascertains that older adults tend to
be emotionally driven and recall personal experiences, when time is perceived

as limited (Carstensen, Fung, & Charles, 2003; Cole et al., 2008; Williams & Drolet, 2005). In this sense, long-term interpersonal relationships, values (e.g., giving and helping), friend's behaviors (Cole et al., 2008), and self-relevant information (Drolet et al., 2017) may influence decision-making and purchase behaviors.

Figure 2.2 shows the interdependence between the self and the other that are the basis of social interactions and the dynamics used in communication and media advertising.

The author Cooley, 1992 emphasizes that the relationship between the self and others is determined by a three-step looking glass concept: (1) We imagine how others see us; (2) we imagine how others judge us; and (3) we react accordingly with that imagination.

As shown in Figure 2.2, one of the most crucial knowledge to an individual is self-knowledge, that is "know thyself" that relies on the cycle process of (de) coding both spontaneous and intentional signs, and comparison with

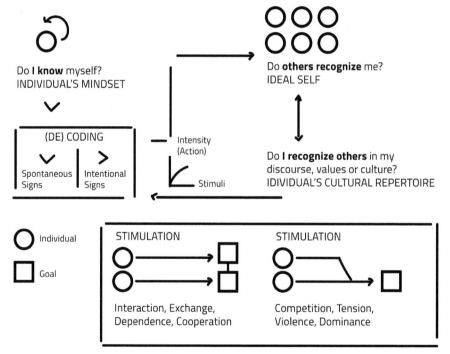

Figure 2.2 Social interaction scheme based on the studies of Bateson (1972), Cooley (1992), and Goffman (2005).

individual's references and cultural repertoire (Bateson, 1972; Cooley, 1992; Goffman, 2005). In case the self and the other have different and complementary goals, scenarios of cooperation, exchange, or dependence interactions are likely to occur. By contrast, competition, tension, violence, and occurrences of dominance or submission tend to prevail when the self and the other have the same goal.

This social interaction process is also transferred to marketing. Indeed, there are six principles that are the basis for influence in marketing (Cialdini, 2014) that derive from self-others' interactions. These are:

- *Reciprocation* refers to both favor-giving and favor-taking, in which there is a need of repaying an action (e.g., gifts, invitations, say yes to a request, free sampling, concessions).
- *Consistency* is relative to a match between promises or deals and real actions/attitudes.
- *Social Cues* refers to the process of mimesis relative to how people feel, think, and act. Indeed, these are the basis of daily interactions.
- *Liking* grounds on the principle of appreciating people who share common hobbies or interests.
- *Authority* principle mentions that people tend to like those who demonstrate expertise in certain area.
- *Scarcity* refers to the fact that people tend to value things that are scarce and, therefore, part of limited offers or one-time deals.

In short, these principles are part of the strategies adopted in different influence strategies-for example, Online Summits, Joint venture webinars, podcasting, among others.

Implementing an influence media advertising plan embodies a considerable number of tasks. In specific, these are:

- Determine the goals, target audience and key performance indicators–KPI[1] (e.g., audience reach, view, engagement, followers).
- Identify the partner influencers whose value/style match the product message and branding.
- Plan a publishing schedule and review the influencer content.
- Monitor engagement.

The following subsections detail each task.

---

[1]Metrics that define a quantifiable measurement that determines the effectiveness and performance of a company, a marketing strategy or campaign with a couple of pre-defined goals.

2.2.1 Determine the Goals, Target Audience and KPI

The first step when establishing an influence media advertising plan is to define the goals to accomplish with the adopted strategy. These goals should be clear (what? why are they important? who are involved? where? which resources?), measurable (track progress-how much/many?), achievable (attainable, success KPI-how to accomplish? is it realistic?), relevant (check whether it is worthwhile to achieve the goal/right time/effort), and time-bound (schedule of when and what to do).

In terms of the target audience, customer personas are usually sketched before driving to sales (Revella, 2015) to target the older adult consumer and (re) think the customer experience (expectation management and influence, purchase, and postpurchase).

A customer persona can be defined as a fictional customer that represents the final consumers (Brangier & Bornet, 2011; Trischler, Zehrer & Westman, 2018). After defining the customer typical purchase behaviors, desires, goals, and context-based scenarios, the necessary product/service requirements are established.

In this sense, it is worth mentioning that different age cohorts should be considered when addressing the older adult consumer. For example, Fisk, and colleagues (2009) mention two groups-that is young-old (60–75) and old-old adults (+75). Lee et al., 2011 divide into young-old (65–74), old-old (75–84), and oldest-old (85+) whereas in our research, we added a younger age group entitled presenior (50–64) to the age cohorts proposed by Lee et al. (2011) and Veloso (2014).

Finally, a set of key performance indicators are identified based on the goals and target audience to assess the effectiveness of the adopted strategy and monitor engagement-for example, number of followers, purchases, conversion rates, and content sharing.

2.2.2 Identify the Partner Influencers Whose Value/Style Match the Product Branding

Word-of-mouth advertising is an effective strategy that influences consumers' purchases. Indeed, recommendations and user-generated content towards branding have a significant impact on purchasing decisions.

Although older adults are a promising segment in online marketing, understanding their Internet usage behaviors is extremely relevant to be effective in the adopted communication strategies (Esteves, Slongo, Barcelos, & Esteves, 2015). Moreover, it seems that negative reviews have more impact on this target group rather than positive ones (Von Helversen, Abramczuk, Kopeć, & Nielek, 2018).

When defining an Influencing Plan for Business, assigning the roles of *mavens*, *connectors* and *salesmen* proposed by Gladwell (as cited in Backaler, 2018) is fundamental to communicate the message. These mentioned roles can be defined as follows:

- *Mavens* are domain experts, who challenge consumers with new ideas. These influencers gain consumers' trust based on the insights and knowledge shared. For mavens, influencing is all about the content to disseminate, and values.
- *Connectors* are the networkers, having an important role in spreading the message. Connectors have particular attention to which channels will be used and rationale, communication frequency, among other aspects.
- *Salesmen* are the negotiators and persuaders, who have a key role in converting prospects into customers. These will assist customers in sales.

These influencer types are interrelated and should not work in isolation. For that, the following actions are important (Brown & Hayes, 2008): idea planting (answering "what if?"), predicting (determining the target group, tendency and a new product), proclaiming (defining which channels to divulgate the message, frequency, and reason), scoping (mapping the limitations and possible solutions), and persuading (determining the strategies to adopt and business deals).

Also, age identity and stereotypes tend to affect communication and media advertisement and as such the following occurrences may cause a distance effect of lessening the sense of self in older adults (Harwood, 2007; Costa & Veloso, 2017):

- Social Mobility occurs when individuals act in accordance with younger age groups. Changing dress style or plastic surgery are some examples.
- Social Creativity occurs when positive aspects relative to ageing are emphasized, for example, association between ageing and wisdom.
- Social Competition occurs when the social image of ageing is challenged. Older adults are compared with younger target groups, demonstrating own value.

In brief, different types of influencers (mavens, connectors, and salesmen) should be intertwined to design media advertising campaigns. When addressing these campaigns to the older target group, there must be special caution to avoid ageing stereotypes.

The next sections cover the next tasks of implementing an influence media advertising plan-that is publishing schedule and monitor engagement.

2.2.3 Plan a Publishing Schedule and Review the Influencer Content

A publishing schedule is relevant to determine which content to be published in which period of time, frequency, and different communication channels. Scheduling is important to create a routine and a sense of connection between influencers and the audience.

In this plan, the type of content, the media channels used, and time frequency should be detailed and adjusted accordingly with engagement metrics.

2.2.4 Monitor Engagement

Finally, monitoring engagement is essential to assess the effectiveness of media advertising campaigns. For that, audience metrics, channels, digital behaviors and influence, reputation, and return on investment-ROI contribute to the success of these campaigns.

Figure 2.3 shows engagement based on the engagement food chain that illustrates the hierarchy relative to the interconnectedness with a product and brand.

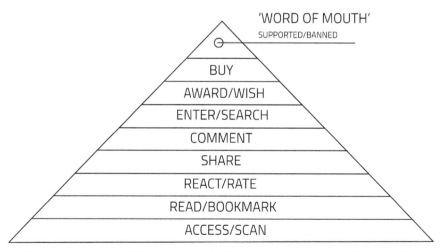

Figure 2.3 Engament Needs based on the Engagement food chain by Sterne (2010).

As shown in Figure 2.3, there are a number of actions that show different levels of compromise with brand/product content from awareness to purchase. In specific, these actions may be grouped into the following:

- Access/Awareness that refers to how the audience know about the company and offerings. Growing members and useful content occur at the basis of the pyramid stage, that is access and navigation to the content, the reading activity and bookmarking.
- Reaction/Registration is relative to the number of people who adhered/subscribed to company services and ensure popular traffic, that is reactions, rates, posts, shares, downloads, and groups formed.
- Commenting/Searching refers to people's engagement in conversations and searching activity, topic interaction, and liveliness, that is comments, mentions, user-generated content, clicks, and messages.
- Awarding/Gifting/Purchasing are relative to the number of buyers from organizations, that is purchases, recommendations, tags, wishlist, and forward to friends

These measures end to have an impact on brand reputation (e.g., trust, responsiveness, relationship building, and recommendations), and ROI (relation between profit/earnings and costs).

In the section that follows, the challenges and opportunities of game market addressed to the older adult segment is addressed.

2.3 The Ageing Process of the Game Market: Challenges and Opportunities

Over the past few years, the game market has shown an increased interest in the older adult consumer. As populations age and a growing number of older adults become regular players (Lenhart, Jones & Macgill, 2008), game-based approaches to maintain cognitive, physical, and social well-being is increasingly important, forming the basis of a growing market.

Previous studies in games for active ageing (Gajadhar, Nap, de Kort & IJsselsteijn, 2010; Nap, De Kort, & IJsselsteijn, 2009) highlight the relevancy of understanding the older adult gamer profile and gameplay activities. Although gameplay in children and youngsters is widely covered, knowledge on gameplay in later adulthood is still in its infancy.

Our previous study in the field revealed that the most popular games in this target group were action/adventure games with problem-solving, strategy and memory games, and card games (Costa & Veloso, 2016). Moreover, these game genres are in accordance with the skills players want to practice. For example, gamers who elected action/adventure games tend to want to practice problem-solving, spatial and temporal memory skills, and calculation. Strategy game-players are more likely to prefer problem-solving and organization.

In terms of the motivations for game-playing in later adulthood, the key role of games as *brain teasers* to keep the mind activity in a fun manner are frequently mentioned. De Schutter & Vanden Abeele (2008) add that players seek games that foster connectedness (the social aspect), "cultivate" their knowledge as well as contribute to society.

Considering that game artifacts reflect many times the socio-cultural context of players (Salen & Zimmerman, 2003) and that wisdom and experience are part of ego integrity in later adulthood (Erikson, as cited in Zastrow & Kirst-Ashman, 2007), game designers are invited to explore this psychosocial development and maturity phase that occurs in the age of 60, in which there is a retrospection of the past lived and a general interest in the topics of memory, narrative and share of personal experiences.

A focus on brain-training games addressed to this target group is also owe to the promising results of the use of games to improve cognitive capacities (Pearce, 2008). Indeed, several lines of evidence (Aison, Davis, Milner, & Targum, 2002; Hall, Chavarria, Maneeratana, Chaney, & Bernhardt, 2012) suggest that video games can improve some cognitive functions such as short-term memory capacity, attention, hand-eye coordination, and autonomy in problem-solving. In this sense, Dr. Ryuta Kawashima, a neuroscience researcher, partnered with Nintendo and such games as Brain Age (Nouchi et al., 2012) and other brain teasing games came up.

The use of games in physical exercising and fall prevention have also been widely covered, grounded on studies that suggest the physical benefits of games-for example, physical rehab and balance improvement. Some of the physical simulation activities are walking soccer, yoga and tai chi (Loos & Zonneveld, 2016), and commercial games include *Kinect Sports*[2], *Wii Fit*[3], and *Dance Dance Revolution*[4].

---

[2]Game published by Microsoft Game Studios for XBOX 360 with Kinect https://market place.xbox.com/pt-PT/Product/Kinect-Sports/66acd000-77fe-1000-9115-d8024d5308c9 (Access Date: Dec 1, 2020).

[3]Wii Fit https://www.nintendo.pt/Jogos/Wii/Wii-Fit-283894.html (Access Date: Dec 1, 2020).

[4]Dance Revolution https://www.playstation.com/en-us/games/dance-dance-revolution-k onamix-psone/ (Access Date: Dec 1, 2020).

The social component of video games for older adults is still to explore, yielding contradictory results. On the one hand, studies (De Schutter & Vanden Abeele, 2008; Ijsselsteijn, Nap, de Kort & Poels, 2007) emphasize the role of games to foster collective interaction and social support. On the other hand, other studies (De Schutter & Vanden Abeele, 2010; Gajadhar, Nap, de Kort & IJsselsteijn, 2010) have pointed out that solitary and single mode games were preferred to this target group in comparison with an online co-playing experience. Understanding this social context associated to the game-playing experience, the use of *Person-versus-Environment* (PvE) and *Person-versus-Person* (PvP) challenges, among other aspects need to be considered.

In all, a change has been observed in the digital game market landscape and designing game products and services catering for the needs and motivations of older adults are more and more in demand (Cota & Ishitani, 2015; Pearce, 2008).

2.4 Concluding Remarks

This chapter discussed the challenges and opportunities posited to the silver market that has been presented as both promising and growing over the past years, given the ageing population and advances in the design and development of gerontechnologies.

Age-neutrality in marketing and the use of technologies in three different domains-that is anti-ageing, assistive/adaptive ageing, and enhanced/ameliorated ageing, have brought to the fore many concerns, such as ethics, technology phobia, cost, accessibility, technology determinism, among others.

In media advertising, influencing is part of the process that may contain a set of principles that bring congruency in self-others' relationships, that is reciprocation, consistency, social cues, liking, authority, and scarcity. For that, an influence media advertising plan helps to address media messages to the target group and when targeting the older adult consumer, different age cohorts, internet usage patterns, and product/service reviews are of particular importance.

Games have also catered for the needs and motivations of the older adult consumer, but much more is still to discover about the cognitive processes associated to games and implications on ageing, while reaching a commercial breakthrough.

References

Aison, C., Davis, G., Milner, J., & Targum, E. (2002). *Appeal and interest of video game use among the elderly* [Working Paper]. Harvard Graduate School of Education.

Backaler, J. (2018). *Digital Influence Unleash the Power of Influencer Marketing to Accelerate Your Global Business*, Cham, Switzerland: Palgrave Macmillan.

Baltes, M. M., & Carstensen, L. L. (1999). Social-psychological theories and their applications to aging: From individual to collective. In V. Bengtson & Schaie (Eds.) *Handbook of theories of aging*, New York: Springer Publishing Company, 209–226.

Bateson, G. (1972). *Steps to an ecology of mind: Collected essays in anthropology, psychiatry, evolution, and epistemology*. Illinois, USA: University of Chicago Press.

Bouma, H., Fozard, J. L., Bouwhuis, D. G., & Taipale, V. (2007). Gerontechnology in perspective. *Gerontechnology*, 6(4), 190–216.

Bouma, H. (1998). Gerontechnology; Emerging Technologies and their Impact on Aging in Society, In J. Graafman et al. (Eds.) *Gerontechnology: A Sustainable Investment in the Future*, Amsterdam, NL: IOSPress, pp. 93–104

Brangier, E., & Bornet, C. (2011). Persona: A method to produce representations focused on consumers' needs. In W. Karwoski, M. Soares, N. Stanton (Eds.) *Human Factors and ergonomics in Consumer Product Design: methods and techniques*, Boca Raton: CRC Press, 37–61.

Brown, D., Hayes, N. (2008). Influencer Marketing-Who Really Influences Your Customer?. *Motivation and emotion*, 27(2), 103–123. https://doi.org/10.1023/A:1024569803230

Carstensen, L. L., Fung, H. H., & Charles, S. T. (2003). Socioemotional selectivity theory and the regulation of emotion in the second half of life. *Motivation and emotion*, 27(2), 103–123. https://doi.org/10.1023/A:1024569803230

Cialdini, R.B. (2014). *Influence Science and Practice*. 5th ed. Harlow, England: Pearson Education Limited.

Cole, C., Laurent, G., Drolet, A., Ebert, J., Gutchess, A., Lambert-Pandraud, R., ... & Peters, E. (2008). Decision making and brand choice by older consumers. *Marketing Letters*, 19(3-4), 355–365. https://doi.org/10.1007/s11002-008-9058-x

Cooley, C. H. (1992). *Human nature and the social order*. London, UK: Transaction Publishers.

Costa L.V., Veloso A.I. (2017) Demystifying Ageing Bias Through Learning. In: Beck D. et al. (eds) Immersive Learning Research Network. iLRN 2017. Communications in Computer and Information Science, vol 725. Springer, Cham. https://doi.org/10.1007/978-3-319-60633-0_17

Costa, L. V., & Veloso, A. I. (2016). Factors influencing the adoption of video games in late adulthood: a survey of older adult gamers. International Journal of Technology and Human Interaction (IJTHI), 12(1), 35–50. DOI: 10.4018/IJTHI.2016010103

Cota, T.T., Ishitani, L. (2015). Motivation and benefits of digital games for the elderly: a systematic literature review. Revista Brasileira de Computação Aplicada, 7(1), p. 2–16. DOI:10.5335/rbca.2015.4190

Czaja, S. J., Boot, W. R., Charness, N., & Rogers, W. A. (2019). Designing for older adults: *Principles and creative human factors approaches*. Boca Raton, FL: CRC press.

Damuleviciene, G., Lesauskaite, V., Knasiene, J., & Macijauskiene, J. (2010). Use of technologies in maintaining autonomy of frail older persons. *Medicina* (Kaunas, Lithuania), 46, 35–42.

De Schutter, B., & Vanden Abeele, V. (2010, September). Designing meaningful play within the psycho-social context of older adults. In *Proceedings of the 3rd International Conference on Fun and Games* (pp. 84–93). https://doi.org/10.1145/1823818.1823827

De Schutter, B., & Vanden Abeele, V. (2008). Meaningful play in elderly life. In *Annual Meeting of the International Communication Association*, Quebec, Canada. Retrieved from https://lirias.kuleuven.be/handle/123456 789/270392 (Access date: May 2nd, 2021)

Drolet, A., Bodapati, A. V., Suppes, P., Rossi, B., & Hochwarter, H. (2017). Habits and free associations: Free your mind but mind your habits. *Journal of the Association for Consumer Research*, 2(3), 293–305. https://doi.org/10.1086/695422

Dziechciaz, M., & Filip, R. (2014). Biological psychological and social determinants of old age: Bio-psycho-social aspects of human aging. *Annals of Agricultural and Environmental Medicine*, 21(4), 835–38. doi: 10.5604/12321966.1129943

Esteves, P. S., Slongo, L. A., Barcelos, R. H., & Esteves, C. S. (2015). Third-Agers on The Internet: Impacts on Word-of-Mouth and Online Purchase Intentions. Procedia economics and finance, 23, 1607–1612. https://doi.org/10.1016/S2212-5671(15)00326-3

Fisk, A. D., Czaja, S. J., Rogers, W. A., Charness, N., & Sharit, J. (2009). Designing for older adults: *Principles and creative human factors approaches*. Boca Raton, FL: CRC press.

Gajadhar, B. J., Nap, H. H., de Kort, Y. A. W., & IJsselsteijn, W. A. (2010). Out of sight, out of mind. Proceedings of the 3rd International Conference on Fun and Games-Fun and Games'10, p.74–83. doi:10.1145/1823818.1823826

Goffman, E. (2005). *Interaction ritual: Essays in face to face behavior*. New Jersey, USA: AldineTransaction.

Grabinski, C. J. (2007). Careers in Aging. In J. E. Birren (Eds.) *Encyclopedia of Gerontology-Age, Aging, and the Ageing*, 2nd ed, p. 230–239

Hall, A. K., Chavarria, E., Maneeratana, V., Chaney, B. H., & Bernhardt, J. M. (2012). Health benefits of digital videogames for older adults: A systematic review of the literature. Games for Health: Research, Development, and Clinical Applications, 1(6), 402–410. doi: 10.1089/g4h.2012.0046.

Hanson, V. L., & Crayne, S. (2005). Personalization of Web browsing: adaptations to meet the needs of older adults. *Universal Access in the Information Society*, 4(1), 46–58. https://doi.org/10.1007/s10209-005-0110-9

Harwood, J. (2007). *Understanding communication and aging: Developing knowledge and awareness*. Thousand Oaks, USA: Sage.

Hawthorn, D. (2000). Possible implications of aging for interface designers. *Interacting with computers*, 12(5), 507–528. DOI: 10.1016/S0953-5438(99)00021-1

Hettich, D., Hattula, S., & Bornemann, T. (2018). Consumer decision-making of older people: a 45-year review. *The Gerontologist*, 58(6), e349-e368. https://doi.org/10.1093/geront/gnx007

Ijsselsteijn, W., Nap, H. H., de Kort, Y., & Poels, K. (2007). Digital game design for elderly users. In *Proceedings of the 2007 conference on Future Play*, New York: ACM, 17–22. doi: 10.1145/1328202.138206

ISO 9241-210:2010. Ergonomics of human-system interaction-Part 210: Human-centred design for interactive systems (n.d.) Retrieved from Jan 13, 2021 https://www.iso.org/obp/ui/#iso:std:iso:9241:-210:ed-1:v1:en

Kohlbacher, F., & Herstatt, C. (2011). Preface and Introduction In Kohlbacher, F., & Herstatt, C. (Eds.). (2011). *The Silver Market Phenomenon: Marketing and Innovation in the Aging Society*, 2nd ed., Heidelberg, Germany: Springer Science & Business Media, p. v-xxii.. https://doi.org/10.1007/978-3-642-14338-0

Lee, P. L., Lan, W., & Yen, T. W. (2011). Aging successfully: A four-factor model. *Educational Gerontology*, 37(3), 210–227. https://doi.org/10.1080/03601277.2010.487759

Lenhart, A., Jones, S., & Rankin Macgill, A. (2008). Pew Internet Project data memo: Adults and video games. Pew Internet & American Life Project, December 7, 2008. Retrieved from https://www.pewinternet.org/wp-content/uploads/sites/9/media/Files/Reports/2008/PIP_Adult_gaming_memo.pdf.pdf

Li, J., Ma, Q., Chan, A. H., & Man, S. S. (2019). Health monitoring through wearable technologies for older adults: Smart wearables acceptance model. *Applied ergonomics*, 75, 162–169. https://doi.org/10.1016/j.apergo.2018.10.006

Lian, J. W., & Yen, D. C. (2014). Online shopping drivers and barriers for older adults: Age and gender differences. *Computers in Human Behavior*, 37, 133–143. https://doi.org/10.1016/j.chb.2014.04.028

Liddicoat S., Newton C. (2019) Older Adults as Co-researchers for Built Environments: Virtual Reality as a Means of Engagement. In: Neves B., Vetere F. (eds) *Ageing and Digital Technology*. Springer, Singapore. https://doi.org/10.1007/978-981-13-3693-5_10

Liu, L., Stroulia, E., Nikolaidis, I., Miguel-Cruz, A., & Rincon, A. R. (2016). Smart homes and home health monitoring technologies for older adults: A systematic review. *International journal of medical informatics*, 91, 44–59. https://doi.org/10.1016/j.ijmedinf.2016.04.007

Loos, E. F. & Zonneveld, A. (2016). Silver Gaming: Serious Fun for Seniors? In J. Zhou & G. Salvendy (Eds.), Human Aspects of IT for the Aged Population. Healthy and Active Aging, Second International Conference, ITAP 2016, Held as Part of HCI International 2016, Toronto, ON, Canada, 17–22 juli 2016, Proceedings, Part II (pp. 330–341). Cham: Springer International Publishing. https://doi.org/10.1007/978-3-319-39949-2_32

Marshall V.W., Bengtson V.L. (2011) Theoretical Perspectives on the Sociology of Aging. In: Settersten R., Angel J. (eds) Handbook of Sociology of Aging. *Handbooks of Sociology and Social Research*. Springer, New York, NY. https://doi.org/10.10Á07/978-1-4419-7374-0_2

Moschis, G. P. (2003). Marketing to older adults: an updated overview of present knowledge and practice. *Journal of Consumer Marketing*, Vol. 20 No. 6, pp. 516–525. https://doi.org/10.1108/07363760310499093

Nap, H. H., De Kort, Y. A. W., & IJsselsteijn, W. A. (2009). Senior gamers: Preferences, motivations and needs. *Gerontechnology*, 8(4), 247–262. doi: 10.4017/gt.2009.08.04.003.00

Nimrod, G. (2012). Online communities as a resource in older adults' tourism. *The Journal of Community Informatics*, 8(1), 1–11.

Nouchi, R., Taki, Y., Takeuchi, H., Hashizume, H., Akitsuki, Y., Shigemune, Y., . . . Yomogida, Y. (2012). *Brain training game improves executive functions and processing speed in the elderly: A randomized controlled trial*. PloS one, 7(1), p. 1–9. https://doi.org/10.1371/journal.pone.0029676

Pearce, C. (2008). The Truth About Baby Boomer Gamers A Study of Over-Forty Computer Game Players. *Games and Culture*, 3(2), 142–174. https://doi.org/10.1177%2F1555412008314132

Pesonen, J., Komppula, R. & Riihinen, A. (2015). Typology of senior travellers as users of tourism information technology. Inf Technol Tourism 15, 233–252. https://doi.org/10.1007/s40558-015-0032-1

Pino, M., Boulay, M., Jouen, F., & Rigaud, A. S. (2015). "Are we ready for robots that care for us?" Attitudes and opinions of older adults toward socially assistive robots. *Frontiers in aging neuroscience*, 7, 1–15. https://doi.org/10.3389/fnagi.2015.00141

Reinmoeller P. (2011) Service Innovation: Towards Designing New Business Models for Aging Societies. In Kohlbacher, F., & Herstatt, C. (Eds.). (2011). *The Silver Market Phenomenon: Marketing and Innovation in the Aging Society*, 2nd ed Heidelberg, Germany: Springer Science & Business Media, p. 133–146.https://doi.org/10.1007/978-3-642-14338-0_10

Revella, A. (2015). *Buyer personas: how to gain insight into your customer's expectations, align your marketing strategies, and win more business*. New Jersey, USA: John Wiley & Sons.

Sale, P. (2018) Gerontechnology, Domotics, and Robotics. In: Masiero S., Carraro U. (eds) *Rehabilitation Medicine for Elderly Patients*. Practical Issues in Geriatrics. Springer, Cham, 190–216. https://doi.org/10.1007/978-3-319-57406-6_19

Salen, K., & Zimmerman, E. (2003). Rules of Play: *Game Design Fundamentals*. Cambridge, Massachusets: MIT Press.

Settersten R.A., Angel J.L. (2011) Trends in the Sociology of Aging: Thirty Year Observations. In: Settersten R., Angel J. (eds) *Handbook of Sociology of Aging. Handbooks of Sociology and Social Research*. Springer, New York, NY. https://doi.org/10.1007/978-1-4419-7374-0_1

Shishehgar, M., Kerr, D., & Blake, J. (2018). A systematic review of research into how robotic technology can help older people. *Smart Health*, 7, 1–18. https://doi.org/10.1016/j.smhl.2018.03.002

Smarr, C. A., Fausset, C. B., & Rogers, W. A. (2011). Understanding the potential for robot assistance for older adults in the home environment

[Technical Report HFA-TR-1102]. School of Psychology, Georgia Institute of Technology. Retrieved from https://smartech.gatech.edu/bitstream/hand le/1853/39670/HFA-TR-1102-RobotSupportForHomeTasks.pdf

Sterne, J. (2010). Social Media Metrics How to Measure and Optimize your Marketing Investment. Hoboken, New Jersey: Sebastopol, CA: Wiley John Wiley & Sons, Inc.

Stroud, D. (2005). *The 50-Plus Market: Why the future is Age Neutral when it comes to marketing & branding strategies.* London and Sterling, VA: Kogan Page.

Trischler, J., Zehrer, A., & Westman, J. (2018). A designerly way of analyzing the customer experience. *Journal of Services Marketing.* Vol. 32 No. 7, pp. 805–819. https://doi.org/10.1108/JSM-04-2017-0138

Vale Costa L., Veloso A.I., Loos E. (2019) Age Stereotyping in the Game Context: Introducing the Game-Ageism and Age-Gameism Phenomena. In: Zhou J., Salvendy G. (eds) Human Aspects of IT for the Aged Population. Social Media, Games and Assistive Environments. HCII 2019. Lecture Notes in Computer Science, vol 11593. Springer, Cham. https://doi.org/10.1007/978-3-030-22015-0_19

VandeWeerd, C., Yalcin, A., Aden-Buie, G., Wang, Y., Roberts, M., Mahser, N., ... & Fabiano, D. (2020). HomeSense: Design of an ambient home health and wellness monitoring platform for older adults. *Health and Technology*, 10(5), 1291–1309. https://doi.org/10.1007/s12553-019-00404-6

Veloso, A. I., & Costa, L. (2014). Jogos na comunidade miOne. In A.I. Veloso (Ed.), *SEDUCE Utilização da comunicação e da informação em ecologias web pelo cidadão sénior.* Porto, PT: Ed.Afrontamento

Von Helversen, B., Abramczuk, K., Kopeć, W., & Nielek, R. (2018). Influence of consumer reviews on online purchasing decisions in older and younger adults. *Decision Support Systems*, 113, 1–10. https://doi.org/10.1016/j.dss.2018.05.006

Williams, P., & Drolet, A. (2005). Age-related differences in responses to emotional advertisements. *Journal of consumer research*, 32(3), 343–354. https://doi.org/10.1086/497545

Zastrow, C., & Kirst-Ashman, K. K. (2009). *Understanding human behavior and the social environment.* CengageBrain. com.

3

Designing Game-based
Tools for Active Ageing

This chapter examines a set of design considerations to develop games for active ageing. It first gives a brief overview of both opportunities and challenges of ageing in a network society, discussing information literacy, and digital inclusion without overlooking the fallacy of technological determinism. The role of design in people's well-being and quality of life is then acknowledged, overstating the aspects to be considered in assistive and wearable e-health platforms. Finally, this chapter ends with the application of games in cognitive and physical rehabilitation, relying on game accessibility.

3.1 Ageing Actively in Digitally-Mediated Spaces

In the light of the proliferation of digital devices in a global society, such challenges as an asymmetric access to information and daily-life services, infocommunication overload and noninclusive interfaces have been brought to the fore as a concern (Holling, 2001). Moreover, a locked-down society has amplified these challenges and the need to address the older adults' context of use and needs (e.g., access to health information, internet-based social support, online banking services, and shopping; Seifert, 2020).

Although there has been increasing interest in designing market-oriented products and services to an aging population, the topics of digital inclusion, literacy, and technological determinism remain often overlooked and of importance to advance knowledge in gerontechnology.

According to the World Health Organization (2002, p. 12), active ageing is "the process of optimizing opportunities for health, participation, and security in order to enhance the older adults' quality of life as people age."

In 2010, the definition was refined to integrate the idea of lifelong learning (International Longevity Centre Brazil, 2015).

Although there has been a boom of market-oriented products and services entitled "technology for active ageing" (as discussed in Chapter 2), there has been a focus on the health-related dimension, overlooking the other pillars of active ageing as security and participation in society (WHO, 2002; Felsted & Wright, 2014).

In ageing actively in digitally-mediated spaces, the following topics are covered: ageing in a network society, information literacy and digital inclusion, and challenges of lifestyle-monitoring technologies, and technological determinism.

3.1.1 Ageing in a Network Society

In the Information and Communication Society, networks have been reconfiguring the way individuals communicate, interact, and perform daily-life activities (Castells, 2001).

Recently, a high dependency on information and communication technologies to overpass the boundaries of social confinement owing to a pandemic influenza have amplified both the advantages and barriers associated to media, including reinforcement of social interactions, expansion of geographic and temporal boundaries, intergenerational gaps in the access of information, changes in communication and authority roles, accessibility, social digital divide, among others (Costa & Veloso, 2016; Fisk, Rogers, Charness, et al. 2009; Sixsmith and Gutman, 2013).

Information and communication technologies have also a crucial role in overcoming these challenges by facilitating the meaning attributed to information and, thus, encouraging prosocial behaviors, the sense of social connectedness and of purposefulness (Costa, 2013; Cabrer & Malanowski, 2009).

Nevertheless, a general disengagement of older adults toward the use of technology has been likely to affect integration and daily interactions within the Information and Communication Society. Possible reasons for this fact are the lack of access to technologies, perception of the need to use digitally mediated devices, and difficulties in its usage given age-related bio-psycho-social changes (Fisk, Rogers, Charness, Czaja, & Sharit, 2009).

Another possible reason is the fact that the development of many market-oriented products or services for active ageing rely on the designers' assumption of the older adult user's cognitive models with the risk of falling under ageism and some bias instead of involving the users in the ideation and design process (Vines, Pritchard, Wright, Olivier, & Brittain, 2015).

Lastly, information literacy and digital inclusion are other aspects to be considered, especially when designing the interfaces and interaction modes with these products.

3.1.2 Informational Literacy and Digital Inclusion

When making informed choices in daily life, literacy, and numeracy skills are essential. The definition of literacy has evolved over the years-from basic reading and writing skills to the capacity of searching, interpreting, and manipulating information (Benavente, Rosa, Costa, & Ávila, 1996; Olson & Torrance, 2009; Webber & Johnston, 2000).

According to UNESCO (2005, p. 21), literacy can be defined as the following:

> *"(. . .) ability to identify, interpret, create, communicate and compute using printed and written materials associated with varying contexts. Literacy involves a continuum of learning in enabling individuals to achieve his or her goals, develop his or her potentials, and participate in the community and wider society."*

Media can constitute "neo-literate" environments considering that signs and meanings are produced and circulated in society and these may facilitate this process of attributing meaning to the information distributed (UNESCO, 2011). Indeed, there are a number of characteristics of these environments. These are enlisted below:

- Transference of literacy skills to daily life activities.
- Individual's participation and usage of literacy skills in environmental interactions through, for example, the delivery of narratives, challenge-based learning, and information on demand.
- A sense of control over the acquisition of literacy skills either in terms of content, time and mode of using.
- Reinforcement of intrinsic rewards and participation opportunities in the community.

It is worth noting, however, that alphabetism differs from literacy. Whereas the first concept refers to the ability to read and write, literacy refers to the cognitive processes that imply filtering, interpreting and transferring the information to other contexts (Webber & Johnston, 2000).

There are many proposed literacy frameworks (e.g., ACRL Literacy Competence Standards, SCONUL-Seven Pillars of Information Literacy, ANZIL-Australian and New Zealand Information Literacy) (Bruce, 2004). However, both the process of coding and decoding information and the mentioned differences between alphabetism and literacy are not widely covered.

Figure 3.1 illustrates the process for coding and decoding information and the way scenarios of informational literacy or digital inclusion can occur.

As illustrated in Figure 3.1, the process of (de)coding information rely on tactile, auditory, visual, and olfactory sensations and use of these in everyday discourse (texts) that can vary in terms of structure (i.e., expository, procedural, transactional, or persuasive; Thwaites, Davis, & Mules, 2002) and type (i.e., physical and digital).

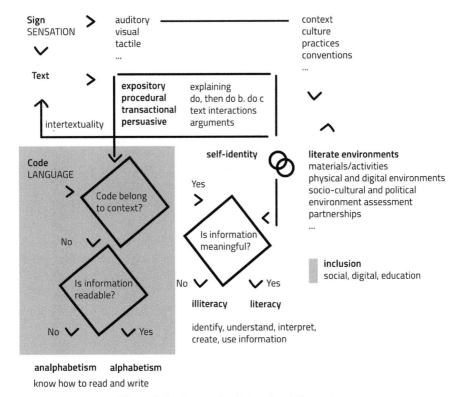

Figure 3.1   Process for (de) coding information.

When text is coded and decoded, a verifying condition whether the code used belongs to the individual's context follows and, as such, the individuals can be considered literate when they can understand, interpret or/and apply the information presented. By contrast, a scenario of alphabetism or digital inclusion refer to the solely process of being able to read the text provided and more complex cognitive processes toward the information are necessary to advance to a scenario of literacy.

In a nutshell, digital inclusion and literacy are essential to bridge inter-generational gaps relative to the use of media, and in particular, games. These neo-literate environments should consider (UNESCO, 2011, p. 10): content and literacy activities, phygital (physical and digital) environments; socio-cultural environments; assessment terms; and collaborative partnerships.

3.1.3 Challenges of Lifestyle-Monitoring Technologies and Technological Determinism

Lifestyle-monitoring technologies for active ageing pose many challenges relative to its usage–for example, ethics, data protection and privacy, literacy, *ageism, accessibility*, among others. Figure 3.2 illustrates some of these challenges.

Figure 3.2   Challenges posed with the usage of lifestyle-monitoring technologies.

In times of social confinement, the following challenges that were the most observed in our applied research involving older adults in the design and evaluation process of digital platforms were the following:

- Informatization of the services and subsequent intergenerational inequality access to those services (e.g., online shopping and electronic payments).
- Access to information and communication technologies, dependency on communication services and info-exclusion.
- Necessity of self-expression and data privacy.

When directing the designed systems to the target group and doing research in digital platforms, technological determinism must be taken into account (Smith & Marx, 1994).

Technological determinism refers to the belief that technology drives society and environment context, usually highlighting possible benefits of technology.

A pioneer in the usage of the term "determinism" was Leibniz to discuss about predictability and fate in the Principle of Sufficient Reason (Earman, 1986) and since then, this concept has extended from this philosophical perspective to refer to influencing events involving different actions and agents.

Alongside technological determinism, there are also other examples of determinism, such as economic (the influence of the way products and services are produced and traded exert in society), environmental (influence of the environment in society), and social (influence of society in the individual). These can be added to the technology usage and be amplified and lead to some inevitable bias (e.g., association of media to violence, addiction, social connection, and information knowledge).

In the perspective of the authors of this book, determinism occurs when a solely attribution of a positive or negative effect is pointed as a cause, without considering other possible effects. As Postman (2011, p. 11) argues:

> "Technological change is neither additive nor subtractive. It is ecological. I mean "ecological" in the same sense as the word is used by environmental scientists. One significant change generates total change."

In view of what has been mentioned so far, criticism is acknowledged when addressing certain technology to a target group and both pros and cons must be balanced.

3.2 Designing for Good, Well-Being, and Quality of Life

We have been witnessing an interesting challenge for the "human-computer interaction" field. In the past, the focus was on the ergonomics of a product and the user was seen as a mere cognitive-physical processor with such indicators as effectiveness, efficiency, and satisfaction (ISO 9241-11:1998, 1998). Then, the role of the human-computer interaction specialist has changed to the crafter of user experience ("what feels like using that product"; Alben, 1996), and, lately, a designer for digital well-being and quality of life has been in need (Calvo and Peters, 2014).

Although some cautions must be applied when adopting human-centered design (e.g., designing for the "today" or "tomorrow" individual, much attention to the users' needs, among others) (Norman, 2005), we believe that "listening" the users' context when designing such systems are an added-value.

In this chapter, the following topics are covered: human-centered design: "What a Wonderful World?", sketching digitally mediated positive experiences, and behavioral design.

3.2.1 Human-Centered Design: "What a Wonderful World?"

The move of human-computer interaction toward more humanistic and natural interactions is not new. Indeed, increasing attention has been drawn lately to the fields of Affective Computing (Picard & Picard, 1997), Social Robotics (Breazeal, Dautenhahn, & Kanda, 2016), and Embodied Conversational Agents (Cassell, 2000).

The popularity of home assistants (e.g., the market solutions Amazon Alexa[1] and First Home Olly[2]) and developments in artificial intelligence (AI) and sentimental computing analysis have led to the increase in emotion recognition and reaction to the users' context (Khan & Lawo, 2016). There are also some drawbacks of the usage of these home assistants such as gathering personal data or increasing isolation with the interaction of these AI-driven personalized agents and simulation of human experiences that combine both virtuality and reality (Best & Lebiere, 2006; Gilbert & Conte, 1995).

---

[1] Amazon Alexa is a virtual assistant that can react to the user's interaction through voice by, for example, playing music or audiobooks, making to-do lists, setting alarms and providing real-time information.

[2] Olly is a home-robot that assists end-users' routines and habits.

Empathy is key to design emotional and affective experiences (Ng, Khong, & Thwaites, 2012; Norman, 2004; Picard & Picard, 1997; Zvacek, 1991). It is worth noting that emotions differ from feelings and according to Damásio (2003) emotions are visible "states of being," actions or movements to others (e.g., tone of voice and expressions) whereas feelings are relative to a perceived image of the body or mind (in case of affective feeling) perturbed by emotions. According to Jordan, 2000, there are four types of pleasures that may be explored to bring a much more humanistic approach to the system. These are:

- *Physio-pleasure* is relative to visual (e.g., graphics, animations), auditory (speech), gustatory, olfactory, and somatic perception (e.g., kinesis, tactile, smell, and flavor) that result from different environment stimuli.
- *Socio-pleasure* refers to social identity (e.g., status, codes, and conventions) and social relationships.
- *Psycho-pleasure* is relative to cognitive-emotional reactions (e.g., moods, dreams, aspirations).
- *Ideo-pleasure* is about personal ideologies and view of products as art.

In Emotional Design, Donald Norman, 2004 lists different types of levels of emotional design:

- *Visceral* is relative to the artefact appearance and aesthetics, inducing attraction or repulsion through an immediate judgement of human senses.
- *Reflective* refers to self-image conscious, memories, personal satisfaction, and sense of accomplishment.
- *Behavioral* is about pleasure and use experience with a possible impact on the repetition of (inter)actions with the system.

In brief, these different types of pleasure and levels of emotional design are important to embody more and more socio-technical dynamics and incorporate community-centred design.

3.2.2 Sketching Digitally Mediated Positive Experiences

Over the past years, the human-computer interaction field has been challenged with a much more humanistic approach and as such new challenges and opportunities have brought to the fore the need to advance on the parameters to design and assess interaction design-that is move from usability parameters to incorporate experience design and digital well-being. These latter advancements are also owing to the emergence of the increasing interest in such topics as Gamification, Games for a Purpose/Change, UX for Good, Behavioral Technology, Wisdom 2.0, and Smart Living.

Currently, a pandemic influenza has also increased the need to find digitally mediated strategies to overpass the boundaries of social distancing measures and have a socio-techno-ethical role in assisting the end-user with a sense of purpose, neighborhood and "a care for place" culture. Although psychological and social metrics (e.g., Torres, 2011; Ferreira et al., 2014; Siegel and Dorner, 2017) have been used to assess the effectiveness of information and communication technologies, they tend to be subjective without considering digitally mediated contextual aspects.

In the book "Positive Computing," Calvo and Peters (2014) highlight that digitally mediated artifacts may impact human potential and well-being, fostering positive emotions and "remembering experiences." According to the authors (2014), the following principles should be followed: (i) encourage autonomy, competence, and self-actualization and (ii) create self-awareness and self-compassion. For these principles, the following procedures are necessary: Foster meta-cognition (i.e., awareness of the thinking process); take into account experimental cognition (e.g., the use of mimesis and do-it-yourself strategies), affect and behavior; invest in mindfulness interventions; and reinforce empathy, compassion and altruism through, for example, storytelling, role-playing and communal goals.

In the specific case of games, they may play a key role in health-related well-being and quality life by taking the following design elements into account:

- Metamemory: Enable players to reflect on their own memories.
- Immediate feedback: Provide immediate feedback relative to action-behaviors and encourage its repetition.
- Context-aware challenges: Adapt the game challenges to the context.
- Storytelling and narrative immersion: Give a purpose of action and behavior change through the use of storytelling.
- Bios, avatars, and role-playing: Provide information about the bios and places visited in a game to provide meaning to the exploration of a world and the player experience.
- Game space: Bring people together in a phygital space (both physical and digital space).
- Imagery-based techniques: Recreate sensory perception of different stimuli.
- Social engagement, social graphs, and community of practice: Embed social media and reinforce the interconnection of online relationships.

3.2.3 Behavioral Design

The integration of digitally mediated solutions in people's daily life activities and the humanistic approach brought to the human-computer interaction field have also led to an increased interest in behavioral design, that is encourage behavior change associated with the use of an artefact in a certain context (Cooper, Reimann, Cronin, & Noessel, 2014).

Media have played a key role in monitoring and "listening" users' behaviors and assist in decision-making. Table 3.1 presents some of the implications of behavior change theories on media design based on the studies of Glanz, Rimer, & Viswanath (2008), Michie, Van Stralen, & West (2011), and Wendel (2013).

Game elements and techniques have a great potential to foster behavior change by, for example, providing positive stimulus or actions and positive feedback in a short period of time. Consequently, players are encouraged to repeat the same actions over and over again and generate a habit.

Table 3.1   Implications for designing media based on behavioral change

Behavioral change theories	Implications in media design
Health belief model: Health behaviors consider both the seriousness of a problem and potential benefits of behavioral change.	Use media design to inform about potential behavior changes and simulation to sensitize the user on the seriousness of a problem and subsequent need of change.
Transtheoretical model: Behavioral change implies five phases (i.e., (1) pre-contemplation: not being interested in behavior change and problem denial; (2) contemplation: a behavior change mindset is not translated into practice; (3) preparation: objectives and plans relative to behavior change are made to the next days; (4) action: behaviors are changed but a routine is still needed; and (5) maintenance: new exercises and intensity changes are necessary to maintain changes in behavior).	Enable scaffolding and level design (e.g., progression to mastery). Divide challenges into "step-by-step" and "smal wins" Diversify the type of challenges/quests and intensity to ensure maintenance in behavior change. Rely on social contacts (e.g., social graphs) to encourage users to behavior change and establish routines.
Relapse: Cognitive and behavior coping strategies minimize lapses in behavior change.	Reward goal achievement and changes in behaviors. Present the user with "recuerdos" of previous achievements and provide alternative coping strategies.

Table 3.1 Continued

Information Processing: Information is built on previous knowledge and experiences, which are coded and retrieved based on different types of visual and auditorial cues, among others.	Associate different types of visual, auditory or tactile cues to daily habits/actions/routines. Easy-to-remember information also facilitates behavior change.
Information Processing: Information is built on previous knowledge and experiences, which are coded and retrieved based on different types of visual and auditorial cues, among others.	Associate different types of visual, auditory or tactile cues to daily habits/actions/routines. Easy-to-remember information also facilitates behavior change.
S*ocial Cognitive Theory*: Individuals are more willing to change their behaviors (self-efficacy) when they believe in their capabilities.	Remind users of prior successes ("small wins") and establish a connection between previous knowledge/past experiences and current status.
Theory of Reasoned Action: "Social pressure" and individual's intentions tend to also affect behavior change.	Promote social accountability and make challenges/actions socially rewarding (e.g., invitation from friends, group events ...)
Community Organization Model: Social communities may impact individual's health, well-being, and behavior change. Social relationships may also influence individual's health behaviors and well-being.	Community building enable trust networks and social status towards behavior change (e.g., ranking). Social modelling (i.e., providing examples for imitating groups) may be also essential to behavior change.
Ecological Models: Healthful choices and behavior change are also dependent on the environment and policy-making.	Provide context-aware information and policies relative to behavior changes.
Organizational Changes: Institutions may also influence and create awareness to certain health-related behavior changes	Encourage long-term relations with the institutions relative to behavior changes through digitally mediated communication and health awareness groups.
Diffusion Innovation: Disseminate innovative health programs to encourage behavior change.	Use media to spread the message relative to changes in behavior.

In brief, media design may foster behavior change by meeting the following implications: (a) Reward goal achievement and provide alternative coping; (b) Associate different types of cues and easy-to-remember information to daily habits and routines; (c) Remind "small wins"; and (d) Reinforce social interactions and context-aware information in behavior change to maintain those changes in daily routines.

3.3 Designing Technologies for an Ageing Population

The increasingly demands of the Information and Communication Society often lead to many socio-economic challenges, in which digital inclusion and generation equality is not frequently ensured. Developing age-friendly environments that assist the everydayness of older adults, city mobility and intergenerational learning is, therefore, essential.

In designing technologies for an ageing population, the following topics are covered: assistive technology (AT) and independent living, wearable technology and mobility, and age-friendly cities and intergenerational learning.

3.3.1 Assistive Technology (AT) and Independent Living

A key aspect of older adults' independent living and regained confidence in daily-life routines in the digital network society is the use of assistive technologies. Indeed, these technologies and games, in particular, may facilitate their access to healthcare and well-being services (Griffiths, Kuss & de Gortari, 2013).

The Assistive Technology Act of, 2004 (29U.S.C. Sec 2202[2]) defines AT as:

> "Any item, piece of equipment, or product system, whether acquired commercially, modified or customized, that is used to increase, maintain, or improve the functional capabilities of individuals with disabilities."

In this mentioned definition, the following characteristics of AT may be extracted: (a) be customizable (adaptive) and (b) increase, maintain (assist) or improve (rehabilitate) the individuals' capabilities. For that, designing and assessing the effectiveness of the designed products/services in terms of individual's functional capacities (i.e., cognitive, emotional, social, and physical) and link to daily-life activities is, therefore, crucial.

For each dimension of these functional capacities and link to daily-life activities, the recommendations based on our experience in the field and literature review (Veloso & Costa, 2016) are detailed below.

(a) *Cognitive dimension*

The cognitive dimension refers to cognition training, logic, creativity, and learning challenges. Some aspects the cognitive dimensions that AT environments can enable are: (a) self-learning (Connolly, Stansfield, & Hainey, 2008;

Gavriushenko, Karilainen, & Kankaanranta, 2015; Ibrahim & Jaafar, 2009); (b) take the users' cognitive condition into account (Evemsen, 2009; Marin, Lawrence, Navarro, & Sax, 20011; Navarro, Lawrence, Marin, & Sax, 2011); (c) attend the users' needs and skills (Alfadhli & Alsumait, 2015; Pinelle, Wong & Stach, 2008); and (d) Adapt the media to the users' activity progress (Alfadhli & Alsumait, 2015; Korhonen & Koivisto, 2006).

Relative to the design recommendations about the cognitive dimension, these are suggested:

- Encourage user/player-tutor interactions (Gavriushenko, Karilainen, & Kankaanranta, 2015).
- Enable a self-learning environment, offering different learning paths (Alfadhli & Alsumait, 2015; Connolly, Stansfield, & Hainey, 2008; Gavriushenko, Karilainen, & Kankaanranta, 2015; Evensen, 2009; Ibrahim & Jaafar, 2009).
- Match the learning content and challenges to the user/players' needs, learning outcomes and learning styles (Alfadhli & Alsumait, 2015; Korhonen & Koivisto, 2006; Law et al., 2008).
- Provide quantitative and qualitative learning assessments (Alfadhli & Alsumait, 2015);
- Divide well-structured information into modules/small chunks (Alfadhli & Alsumait, 2015).
- Foster game and other media-based learning events and provide a fun experience (Wærstad & Omholt, 2013; Torrente, Freire, Moreno-Ger, & Fernández-Manjón, 2015).

(b) *Emotional dimension*

The emotion dimension is relative to the way games and other digital media stir emotions with the use of sound or visual effects (de Oliveira Santos, Ishitani & Nobre, 2013; Federoff, 2002; Papaloukas, Stoli, Patriarcheas, & Xenos, 2010), developing a sense of empathy (Alfadhli & Alsumait, 2015; Desurvire, Caplan, & Toth, 2004; Gavriushenko, Karilainen, & Kankaan-ranta, 2015) and users' emotionally and viscerally involvement (Alfadhli & Alsumait, 2015; Desurvire, Caplan, & Toth, 2004; Glinert, 2008).

In terms of the design recommendations about the emotional dimension, these are suggested:

- Reinforce the emotional connection between users/players and the system/game world (Nackle et al., 2009).

- Immerse the user/player in an emotionally and viscerally experience (Alfadhli & Alsumait, 2015; Desurvire, Caplan, & Toth, 2004; Glinert, 2008).
- Provide different stimuli from different sources and stir emotions through sound and visual effects (Desurvire, Caplan, & Toth, 2004; Federoff, 2002; Papaloukas, Stoli, Patriarcheas, & Xenos, 2010; Pinelle, Wong & Stach, 2008; Wærstad & Omholt, 2013).
- Enable users/players' self-expression through sharing own content (Connolly, Stansfield, & Hainey, 2008; Korhonen & Koivisto, 2006; Paavilainen, 2010).
- Foster empathic interactions between users/players and virtual agents/avatars (Alfadhli & Alsumait, 2015; Desurvire, Caplan, & Toth, 2004; Gavriushenko, Karilainen, & Kankaanranta, 2015; Parnell, Berthouze, & Brumby, 2009).

(c) *Physical dimension*

In the physical dimension, designers must attend the users' physical conditions and create goals that meet the users/players context. The recommendations are the following:

- Incorporate physical exercises training muscle strength and power to prevent the risk of from falling (Santos & Knijnik, 2009).
- Assess health improvements (e.g., clinical test for assessing fall risk; Santos & Knijnik, 2009).
- Guide users/players to perform different motion patterns when they are not familiar with new gestures (Marinelli & Rogers, 2014).
- Adopt the physical challenges to the user/player's conditions and context (Law & Sun, 2012; Sáenz-de-Urturi, García Zapirain & Méndez Zorrilla, 2015; Paavilainen et al., 2009; Sánchez, Vela, Simarro, & Padilla-Zea, 2012).

(d) *Social dimension*

The social dimension refers to users/players' social interactions and social-based challenges (competition and cooperation) (Connolly, Stansfield, & Hainey, 2018; Federoff, 2002; Paavilainen, 2010; Veloso & Costa, 2015), self-expression and communication (Alfadhi & Alsumait, 2015; Connolly, Stansfield, & Hainey, 2008; Paavilainen, 2010; Papaloukas, Patriarcheas, & Xenos, 2009; Sánchez, Vela, Simarro, & Padilla-Zea, 2012).

Some of the recommendations on the social dimension are:

- Enable users/players to create/manage an identity in the digital environment (Sánchez et al., 2012).
- Support both competition and cooperation-based challenges (i.e., challenges, puzzles, characters; Aghabeigi, 2011).
- Encourage the formation of communities.
- Enable self-expression and social contacts (Connolly, Stansfield, & Hainey, 2008: Korhonen & Koivisto, 2006).
- Provide users/players means to share information, resources, and stories (Paavilainen, 2010).
- Foster mobile and web-based interactions (Paavilainen, 2010).

Beyond the recommendations listed in each dimension of the individual's functional capacities to be encouraged with the use Assistive Technologies, the interconnectedness with daily-life activities. This interconnected may be fostered through the relatedness of storytelling experiences to physical events that hook the individuals' interests.

The next section describes the way technology can assist the individuals in the context of urban mobility and ageing-in-place with independence and safety.

3.3.2 Wearable Technology and Mobility

Wearable technology has been very popular these days, being applied in a number of different contexts, that is rehabilitation, fitness, tourism, among others. When applied to the healthcare and lifestyle market in an increasingly ageing society, wearables may encourage aging-in-place with safety and assist older adults' daily habits. Another context in which wearables can contribute to active ageing is facilitating urban mobility.

Over than forty percent of older adult population in the countries of the Organization for Economic Co-operation and Development (OECD) live in urban areas (OECD, 2015). Although this fact may suggest an indicator of business drivers and human development, traffic congestion and lack of accessibility constitute some of the challenges faced daily by older population with a great negative impact not only on city annual costs but also general well-being and quality of life.[3]

---

[3] According to the European Urban Mobility Report (European Union, 2017), the estimated total costs in EU, Norway, and Switzerland overpass 200 billion and most of the pedestrian's fatalities in EU urban areas involve older adults.

In the light of this problem, this section shares some of the lessons learnt from the study and design of a gamified digital app entitled SeriousGiggle that addresses education for mobility. In this digital app for smartwatch or mobile phone, a set of missions and routes are customized to the users' context and mobility condition, relying on a multipeer review system.

Figure 3.3 shows the app screens that proposes a gamified multipeer review system relative to different places, fostering emotional and social connections with cities and the shareability of real-time data about the place or/and routes, that is pavement conditions, wayfinding, route guidance, walking assistance. Seriousgiggle embodies four main functionalities-MyRoute, TravelFit, Missions, and Progress. These are described as follows:

- *MyRoute* aims to guide the user in their walking activities based on context-based information and location. Information about local traffic signs, means of transport, street conditions are examples of the information retrieved from passers-by and travelers' shared twitters and other social media;
- *Missions* consists of challenges incited by friends, caregivers or the system (daily missions, tweets, or other social media). Users can either accept or refuse these challenges clicking on happy or sad emoticons assigned for each action.
- *TravelFit* integrates access to different routes that are based on the end-users' data to present information about outdoor activities-that is peaceful surroundings, sense of community and crime safety, existence of sidewalks, avoidance of urban traffic, and pavement conditions.
- *Progress* is relative to activity monitoring and progress. For each action (Missions, TravelFit, and MyRoute), users can monitor their progress.

The same digital app was also developed for smartphones, in which users can also add photos associated to the place context (e.g., pavement conditions).

Figure 3.3 shows the app screens for the mobile version.

To ensure its sustainability, information is provided by the community of passers-by and these are motivated by a set of missions based on training and social activities promoted by friends or the city council's tourism entities.

The completion of these missions will help to provide sufficient data to constitute accessible and customizable journey plans with route guidance and/or waling assistance rated in terms of safety, community support, environment, and age-friendliness (Figure 3.4).

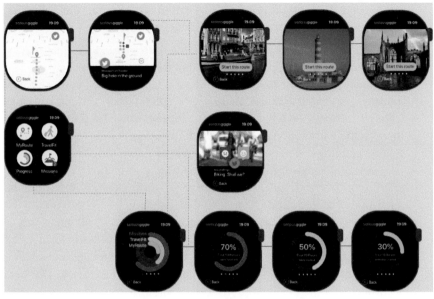

Figure 3.3 App screens (smartwatch).

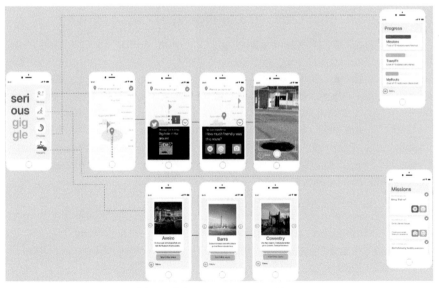

Figure 3.4 App screens (mobile).

In the context of mobility and these gamified systems, the following design recommendations can, therefore, be drawn:

- Add history-based scenarios related to the routes and places that may be also reviewed or added by local communities.
- Personalize a location-based challenges to meet the users' context. Some journey plans may be promoted by city council's or tourism activities (official information), but others may be suggested by the community through the completion of gamification missions.
- Enable activities' schedule and shareability of own progress.
- Use key performance indicators beyond the completion status of missions (e.g., time frequency, accuracy of exercises) to motivate the repetition of actions and formation of daily habits.
- Strengthen social support networks through the use of common activities. In sum, congregating both physical and digital spaces in such a way that neighborhoods are the game interface may foster the connection of different generations with the land and locative narratives. The next section covers the considerations to develop age-friendly environments and its potential in intergenerational learning.

3.3.3 Age-Friendly Cities and Intergenerational Learning

The development of digital technologies for older adults' independent living have been one of the top priorities of the European Commission (2019). Indeed, age-friendly cities should not be regarded as a privilege, but a basic Human Right (World Health Organization, 2002)-the right to the city in its habitat (environment) and habiter (way of living; Lefebvre, 1996).

Information and communication technologies have undoubtedly facilitated social support regardless physical distance (Dewhurst et al., 2015; Leung & Lee, 2005; Whyte & Marlow, 1999; Wright, 2000) and the formation of social networks which may be of social value and a driver force for a change in the neighborhood and city communities.

Our experience in the field of ageing studies, psychology, human-computer interaction, and games have revealed the following aspects for generating an age-friendly environment (Costa, Veloso, Loizou, & Arnab, 2018; Vale Costa et al., 2018):

- Take into account different age-cohorts when addressing products to the target group.
- Do not focus on age-related difficulties or illness when communicating a product/service within the silver market.

- Familiarize the participants with the interface.
- Train cognitive and physical skills, fostering life-long learning and daily life activities.
- Offer opportunities where different generations can work together.
- Reinforce the end user's presence in the environment and stimulate their subconscious by overweighting internal motivations (i.e., skills, beliefs, and self-efficacy) and external elements that lead to human behavior.
- Establish an interlink between cognitive and affective dimensions.

Considering that digital media can have a key role in facilitating an age-friendly environment, they can also constitute a challenge by adopting a technocentric approach that overlooks the proximity between citizens and products/services, especially the ones provided by local communities. For that, a "Smart ecosystem" is essential.

According to ASLERD (2016, p. 3), a "Smart ecosystem" is an environment in which "individuals that take part in the local processes have a high level of skills and, at the same time, they are also strongly motivated and engaged by continuous and adequate challenges, provided that their primary needs are reasonably satisfied." Digital games may also contribute to these environments through the offer of playing activities that enable scaffolding/different levels of progression and collection of skills, rewarded motivated actions and the citizens' involvement in challenges/missions that can impact on society.

Another aspect that is essential in the formation of age-friendly physical and digital environments is intergenerational learning. Although the study of intergenerational relationships can be traced back to "The Problem of Generations" in Mannheim's 1923 essay (Pilcher, 1994), information has been still very scarce until the 80s, and recently the nature of this intergenerational relationship in computer-mediated communication has awaken the interest of the scientific community (e.g., Abeele & De Schutter, 2010; Mayasari, Pedell, & Barnes, 2016). These relationships often involve both solidarity and conflicts (Harwood, 2007), having an impact on each generation's well-being and quality of life (Walker, 2002).

Thomas (2009, p. 5 as cited in Findsen & Formosa, 2011, p. 171) ascertained that intergenerational learning can be defined as a result from "activities which purposely involve two or more generations with the aim of generating additional or different benefits to those arising from single generation activities. It generates learning outcomes, but these may or may not be the primary focus of the activity. It involves different generations learning from each other and/or learning together with a tutor or facilitator." In this context, game-based interactions may bring many benefits (De la Hera, Loos, Simons & Blom, 2017; Zhang & Kaufman, 2016), such as:

- Strength of the familial bond through, for example, regular contact and communication.
- Exchange knowledge transmission from one generation to another.
- Change the perception of each generation.
- Promote in-game communal activities.
- Balance the players' skills (young and old generations) in the gameplay activity.

Kenner et al. (2007) draw our attention to a set of intergenerational engaged activities based on their empirical study, which are:

- Storytelling involving family members and history.
- Religious activities.
- Shopping.
- Playing.

In addition, they also emphasize that cultural heritage and value transmission tend to motivate young and old generations (Harwood, 2007; Kornhaber & Woodward, 1981). Voida & Greenberg (2012) add that games may play a key role in intergenerational learning by encouraging pro-social behaviors and peer-to-peer mentoring. For that, game interaction should be balanced to provide an optimal experience (balance the players' skills and challenges; Csikzentmihalyi, 2008) to different generations of players.

Considering that intergenerational learning relies on the connectivity established between different generations, identifying the different scenarios that can occur and factors is essential. Figure 3.5 shows different scenarios and some of the factors that can affect intergenerational interactions.

From the model of intergenerational connections presented in Figure 3.5, the three scenarios that can occur are:

- Strong connectivity (multiple generations together): In this scenario, each generation is an active agent in the interaction process, ensuring trust, attachment and transmission of values, knowledge and practices.
- Triadic encounters (the generation messenger): In this scenario, an intermediary generation acts as a messenger between younger and older generations. Alliances and coalitions may occur;
- Weak connectivity (The generation crisis): In this scenario, intergenerational connections are occasional and deficiencies in the transmission of values and knowledge in daily practices are owe to the lack of contact.

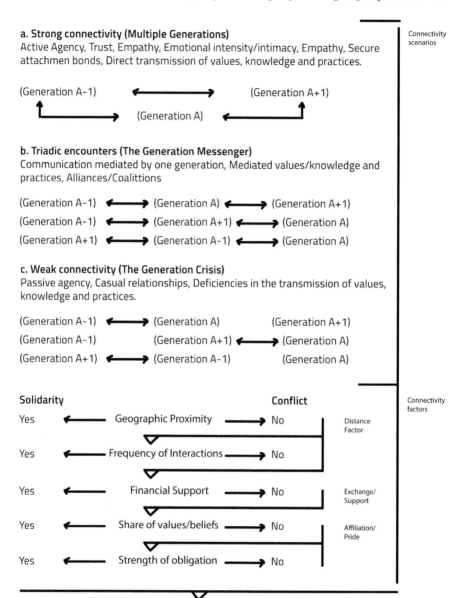

a. Strong connectivity (Multiple Generations)
Active Agency, Trust, Empathy, Emotional intensity/intimacy, Empathy, Secure
attachmen bonds, Direct transmission of values, knowledge and practices.

(Generation A-1)   ⟷   (Generation A+1)

(Generation A)

b. Triadic encounters (The Generation Messenger)
Communication mediated by one generation, Mediated values/knowledge and
practices, Alliances/Coalittions

(Generation A-1) ⟷ (Generation A) ⟷ (Generation A+1)
(Generation A-1) ⟷ (Generation A+1) ⟷ (Generation A)
(Generation A+1) ⟷ (Generation A-1) ⟷ (Generation A)

c. Weak connectivity (The Generation Crisis)
Passive agency, Casual relationships, Deficiencies in the transmission of values,
knowledge and practices.

(Generation A-1) ⟷ (Generation A)   (Generation A+1)
(Generation A-1)   (Generation A+1) ⟷ (Generation A)
(Generation A+1) ⟷ (Generation A-1)   (Generation A)

Connectivity scenarios

Solidarity   **Conflict**

Yes ⟵ Geographic Proximity ⟶ No   Distance Factor

Yes ⟵ Frequency of Interactions ⟶ No

Yes ⟵ Financial Support ⟶ No   Exchange/ Support

Yes ⟵ Share of values/beliefs ⟶ No   Affiliation/ Pride

Yes ⟵ Strength of obligation ⟶ No

Connectivity factors

Individual and Collective Health, Wellbeing, and Quality of Life

Figure 3.5 Model of intergenerational connections.

Relative to the factors that may contribute each scenario, these are grouped into distance, exchange/support, and affiliation/pride (as suggested by the categories proposed by Harwood, 2007). In specific, these factors include:

- Distance factor is related with geographic proximity and subsequent frequency of interactions. Copresence and computer-mediated communication may help to overcome physical distance and inherent obstacles to intergenerational connections;
- Exchange/support factor refers to financial or/and instrumental support. This include creating an appropriate environment for a mutual need to exchange knowledge, values, and beliefs.
- Affiliation/pride factor embodies both the willingness to transmit values and knowledge and the strength of obligation/responsibility to care about the younger/older generation.

In a broader sense, each generation language, self-identity, and context may determine the intergenerational connectivity scenario they may encounter (i.e., strong connectivity, triadic encounters, and weak connectivity) and the factors that may influence those (i.e., distance, exchange/support, and affiliation/pride). Moreover, it is worth noting that context may be influenced by media portrayals, environment, activities, and shared goals/solutions (Kalisch, Coughlin, Ballard, & Lamson, 2013; Kalliopuska, 1994; Schuller, 2010).

In sum, age-friendly cities and intergenerational learning are of importance to ensure co-learning, and a more inclusive and solidary society. Whereas game-based tools can enable cultural transmission and co-learning, caution must be applied to communication accommodation (the adjustment of speech) and some sort of age stereotypes within the context of digital media.

3.4 Cheating Degenerative Diseases through Games

Alzheimer and Parkinson are two examples of brain diseases, in which games can constitute some sort of nonpharmacological interventions. Indeed, games can rehabilitate through cognitive training and increase balance and gait speed in people diagnosed with these diseases. Also, they have been used in rehabilitation and physical exercising, paving the way into home-based training and fitness. For that, game accessibility is essential to design and assess these games. This potential of games in training and coping with diseases and design considerations are, thus, covered in the subheadings: Cheating Alzheimer's and Parkinson's diseases, Stroke rehabilitation, Brain training games and neurorehabilitation, and Exergames and older adult's fitness.

3.4.1 Cheating Alzheimer's and Parkinson's Diseases

According to the World Health Organization (2020), about 50 million of the global population suffer from dementia, being Alzheimer representative of nearly 60–70% of cases. In the same vein, nearly 10 million people have the neurodegenerative Parkinson disease (European Parkinson Disease Association, 2020) and most of the cases also end up developing dementia (van Balkom et al., 2019). Both diseases are of particular concern given that they are aggravated with the ageing process, often leading to patients' institutionalization and caregivers' burden with social and economic costs (Alzheimer's Disease International, 2010; Benveniste, Jouvelot, & Péquignot, 2010; Schrag, Hovris, Morley, Quinn, & Jahanshahi, 2006).

Alzheimer is a progressive and irreversible neurocognitive disease, in which nerve cells die, synaptic connections are damaged, and there is a general shrinkage of the patient's brain (National Institute on Aging, 2019). As a consequence, the following cognitive functions tend to be affected: memory loss, attention deficit, language problems, among others (Kempler, 1995; Morris & Kopelman, 1986; Perry & Hodges, 1999; (Francillette, Boucher, Bouchard, Bouchard, & Gaboury, 2021) with repercussions on patients' behavior, emotions, personality and activities of daily livings (Balsis, Carpenter, & Storandt, 2005; Matyr & Clare, 2012).

There are three progression stages of the Alzheimer's disease. These are (Alzheimer's Disease International, 2013):

- *Earliest or preclinical stage*. In this stage, the disease symptoms are not noticeable. Patients may have some difficulties in retrieving some memories and reveal some time/space confusion or people recognition. These may affect patients' mood, socialization, capacity of dealing with new pieces of information, and daily-life activities.
- *Middle stage* refers to a period in which patients' symptoms are worsened, leading to difficulties in communication and need of assistance to perform daily life activities. In this stage, patients may show apathy or indifference, rejection to care, and some anxiety.
- *Advanced or final stage*: In this stage, patients are dependent on caregivers and need of constant support and surveillance. Patients may experience of some personality changes, delusions, memory loss of episodic events, apathy, or agitation.

The Parkinson neurodegenerative disease results from a deterioration of dopamine-producing neurons, which play an important role in sending electrical signals and relaying messages that determine body movements. As a result, motor problems are likely to occur (namely what is known by the acronym TRAP [Frank, Pari, & Rossiter, 2006]):

• Tremors, for example, trembling hands, fingers.
• Rigidity, for example, arms or legs' stiffness.
• Akinesia that refers to the lack of ability to make voluntary movements. This can include hypokinesia (reduction of movements' amplitude), and bradykinesia (slowness in movement).
• Postural instability and lack of balance.

Beyond these motor impairments, nonmotor problems are also associated Parkinson-for example, sleep disorders, depression, pain, urinary disorders, mood, and attention (Chaudhuri, Healy, & Schapira, 2006; Chaudhuri, Odin, Antonini, & Martinez-Martin, 2011; Kurtis, Rodriguez-Blazquez, Martinez-Martin & ELEP Group, 2013). The speech motor impairment Dysathria may also be also a symptom of Parkinson (Pinto, Ozsancak, Tripoliti, Thobois, Limousin-Dowsey, & Auzou, 2004).

A three-stage based on the Hoen and Yahr stages can characterize the Parkinson disease (Rong, Dahal, Luo, Zhou, Yao, & Zhou, 2019):

• Early-stage: Only one-side symptoms of the body are manifested and evolve to bilateral involvement.
• Mid-stage: In this stage, symptoms on both sides and minimal difficulty walking is manifested (e.g., postural instability).
• Advanced stage: Difficulties in walking may vary from moderate to total inability and use of wheelchair.

Nonmotor aspects of Parkinson based on the Unified Parkinson's Disease Rating Scale (Goetz et al., 2008) also vary from normal, slight-severe in terms of cognitive impairment (i.e., cognitive slowing, impaired reasoning, memory loss, and deficits in attention); hallucinations and psychosis (i.e., illusions, hallucinations, and sensory domains), depressed mood (i.e., low mood and hopelessness), anxious mood (i.e., anxious feelings, tense, and ability to engage in social interactions), apathy, sleep problems, pain, constipation, fatigue, handwriting, among others.

Games have been presented as very promising, having the following benefits:

• Exercise cognitive (e.g., memory, attention, decision-making, visual search) and physical capacities (e.g., De Melo Cerqueira et al., 2020; Lancaster et al., 2020; García-Betances, Arredondo Waldmeyer, Fico, & Cabrera-Umpiér rez, 2015; Nef et al., 2020; Pachoulakis, Xilourgos,

Papadopoulos, & Analyti, 2016; Uğur & Sertel, 2020; Zheng, Chen & Yu, 2017; Yuan et al., 2020).

- Complement patients' treatment, and rehabilitation (e.g., Alves et al., 2018; Muscio, Tiraboschi, Guerra, Defanti, & Frisoni, 2015; Robert et al., 2014).
- Create awareness to these disease (e.g., Cook & Twidle, 2016; Kroma & Lachman, 2018).
- Assist in the diagnosis process or/and cope with daily life activities (e.g., Paletta et al., 2020; Rings et al., 2020).
- Establish a social bond and reinforce intergenerational connections (e.g., Cohen, Firth, Biddle, Lloyd Lewis, & Simmens, 2009; Robert et al., 2014).

In response, the following design recommendations are suggested in the literature for addressing games to these patients:

- Personalize the game experience to the patients' journey (level of disease, abilities, and needs) (e.g., Chauveau, Szilas, Luiu, & Ehrler, 2018; Pachoulakis, Xilourgos, Papadopoulos, & Analyti, 2016);
- Simulate scenarios that patients may encounter in daily routines to help them to cope with daily life activities (e.g., Oña et al., 2020; Szilas et al., 2020; Vallejo et al., 2017);
- Include cognitive challenges-for example, memory games, completing the gaps (e.g., Pilcher, 1994).
- Direct game challenges to the step challenges of the disease-for example, Handwriting, Exergames-Step performance, leg extensions/kicks, and motions (e.g., Dias et al., 2020);
- Reduce the number of stimuli within the game environment (e.g., Francillette, Boucher, Bouchard, Bouchard, & Gaboury, 2021) and promote relaxation through the use of multi-sensory stimulations (e.g., Encalada et al., 2019);
- Provide players' guidance and avoid exploration challenges owe to patients' difficulty in performing self-initiated activities (e.g., Ben-Sadoun, Manera, Alvarez, Sacco, & Robert, 2018);
- Divide a challenge into small number of tasks– "small wins" (e.g., Robert et al., 2014);
- Offer the possibility to extract game data to monitor the patients' evolution with the disease (e.g., Chaldogeridis, Tsiatsos, Gialaouzidis, & Tsolaki, 2014; Schmidt et al., 2020).

In sum, games have shown a potential to impact on Alzheimer and Parkinson's diseases by serving as neuroprotective and neurorestorative exercises. Cognition and physical exercises were pointed out to play a key role in the rehabilitation process and be complementary to traditional rehabilitation programs. The next sections will also cover the application of these games on stroke rehabilitation and the implications of designing brain training challenges and exergames.

3.4.2 Stroke Rehabilitation

Cardiovascular diseases are one of the major causes of death worldwide (Mc Namara, Alzubaidi, & Jackson, 2019), embodying different types of diseases of the circulatory system, for example, ischemic heart disease, and stroke.

Strokes result from the obstruction of the blood supply to part of the brain, often causing brain damage, long-term disability, and even death (Centers for Disease Control and Prevention, 2020a). They can be divided into three categories: Ischemic, Hemorrhagic, and Transient ischemic attacks. Each category is presented below (Centers for Disease Control and Prevention, 2020b):

- Ischemic strokes: These types of strokes are the most common and occur with the partial or total blockage of blood vessels that would supply the brain with oxygen.
- Hemorrhagic strokes: These types of strokes refer to internal bleeding of a brain artery.
- Transient ischemic attack: Blockage of the blood flood in the brain for a short period of time.

As the prevalence and incidence of stroke worldwide increases, cost for treatment and rehabilitation also represents a burden to the health system (Truelsen, Ekman, & Boysen, 2005). Also, an age group who is greatly affected by stroke is the group of senior citizens (Miller, Navar, Roubin, & Oparil, 2006) and some of the stroke effects tend to be aggravated with the ageing process-for example, motor impairments, lack of mobility, difficulties in attention and language, among others (Lodha, Naik, Coombes, & Cauraugh, 2010; Loetscher, Potter, Wong, das Nair, 2019; Pedersen, Vinter, & Olsen, 2004). Three challenges emerge in stroke management, that is ensure the safety, effectiveness, and reduced cost in treatments; deliver stroke-care delivery; and patients' transport to stroke centers (Klijn & Hankey, 2003).

In this sense, investigating games for rehabilitation is, therefore, essential in the process of recovery (Aşın, Atar, Koçyiğit, & Tosun, 2018), and pain distraction (Jameson, Travena, & Swain, 2011; Wiederhold & Wiederhold, 2007). These have already been used in motor rehabilitation and occupational therapy, however, the effects of stroke and implications in game design have not been widely covered to ensure that home-based exercises and movements are not clinically inaccurate.

Ribeiro, Veloso, & Costa (2016) summarize some of the impairments experienced by a patient with stroke based on ICF (International Classification of Functioning) CORE SET (Bickenbach, Cieza, Rauch, & Stucki, 2012) indicating the main implications in game design. Table 3.2 synthetizes the stroke effects and its implications on game design.

As shown in Table 3.2, there are a number of stroke effects that go beyond changes in cognitive processes or motor capacity (e.g., spasticity, muscle atrophy) (e.g., Watkins et al., 2002 to integrate social challenges in dealing with daily-life activities (Bieńkiewicz et al., 2015).

The traditional physical therapy and rehabilitation performed in a hospital context usually involve repetitive-task training (Langhome, Bernhardt, & Kwakkel, 2011) supervised by a health professional. Rehab games can assist this classic rehabilitation process of a patient with stroke (Burke et al., 2009; Lewis, Woods, Rosie, & Mcpherson, 2011), given that the repetition of the same movements tend to be painful and not engaging over time.

The games "A Priest in the Air" (Dias, Veloso, & Ribeiro, 2019) and "Physiofun" (Ribeiro, Veloso, & Costa, 2016) developed under the project SEDUCE 2.0 had the purpose of helping patients to relearn motor skills. These games played either with a bracelet,[4] or/and *leap motion*[5] to retell the story of the medieval era in which the priest *Bartolomeu de Gusmão*, who was mostly famous for presenting one of the first air balloon prototypes (known as *Passarola*) to the Portuguese Court in 1709. The goal of the game is to help the priest to fly the air balloon to the King's castle, deviating from obstacles and catching stars. Figure 3.6 shows the players' hand detection to play the game.

---

[4]This bracelet contains a microcontroller connected to an Inertial Measurement Unit (IMU) to gather patient's data relative to the body's force, angular rate, and magnetic fiels around the body, using an accelerometer, gyroscope, and magnetometer (Ribeiro, Veloso, & Costa, 2016).

[5]The Leap Motion Controller is used for detecting hand gestures and finger points. Ultraleap, retrieved from https://www.ultraleap.com/product/leap-motion-controller/ (Access date: Dec. 30, 2020)

Table 3.2   Stroke effects and implications on game design (Ribeiro, Veloso, & Costa, 2016; Veloso, Costa, & Ribeiro, 2016)

Stroke Effects	Description	Implications in Game Design
Cognitive-Consciousness	The following cognitive functions are affected: Orientation, Intellectual functions, and Higher-level cognitive functions: decision-making, abstract thinking, and task planning (Geyh et al., 2004).	Specific and clear game goals; Use of visual and auditory feedback. Minimalistic design and simplicity in game objects/game environment.
Cognitive-Attention	Decline in selective and divided attention (Spaccavento et al., 2019): loss of focus on different stimulus for a short period time and recall information.	Train attention and problem-solving through the use of games, giving patients more time to execute actions, and provide feedback.
Cognitive-Language	Problems in language functions (Kyeong, Kang, Kyeong, & Kim, 2019). Loss of specific mental function, such as recognizing signs, symbols or idioms.	Avoid voice inputs or auditory outputs without the support of illustrations/simple messages. Stimulate the reading, language comprehension and spelling through the use of game storytelling.
Physics-Vision	Visual impairments can occur (Rowe et al., 2009), such as: Homonymous Hemianopia, Homonymous quadrantanopia, Scotoma.	Specific and clear game goals; Use of visual and auditory feedback. Minimalistic design and simplicity in game objects/game environment.
Physics-Hearing	Peripheral hearing loss (Koohi, 2014) and difficulty in perceiving sounds distorted by noise.	Incorporating subtitles and sound adjustments. Separate ambient noise from character speech.
Physics-Touch and Mobility	Tactile Loss-for example sensing surfaces and texture (Lin, He, Shu & Jia, 2020). Patients' movements tend to be slower and there is a prevalence of spasticity (Watkins et al., 2002). A general difficulty in executing complex movements and little control in performing voluntary movements are also observed.	Games should enable scaffolding from performing simple movements with low amplitude to the ones that are more complex. The movement amplitude and diversity should be gradually introduced in the game;
Social-Participation	Patients are generally challenged with a loss of autonomy in performing everyday tasks (Bieńkiewicz, Brandi, Hughes, Voitl, & Hermsdörfer, 2015). Mood swings are frequent (Cecil et al., 2011).	Foster social relationships and a sense of belonging to a community within a game context.

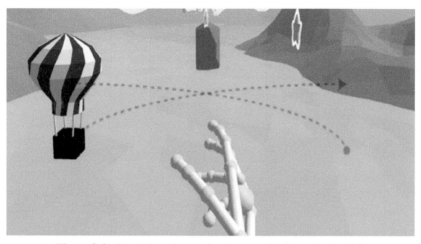

Figure 3.6 Hand detection to play the game "A Priest in the Air."

In terms of gameplay, the player advances in different levels by performing the following actions:

- Lift a balloon through the use of pronation and supination movements. The game informs the player whether the movement was (in)accurately performed and each movement sequence is counted.
- Internal (medial) and external (lateral) rotation of shoulder to deviate from obstacles, and catch stars by positioning the balloon to left, center, and right.

In sum, digital games may be complementary to rehabilitation therapy programs and used either in physiotherapy centers or patients' in-home trainings. The next section describes its specific use in brain training and neurorehabilitation.

3.4.3 Brain Training Games and Neurorehabilitation

During the ageing process, functional and chemical changes are likely to occur in the individual's brain (Goh & Park, 2009), which functions like a muscle endowed with plasticity. If, on the one hand, ageing has been associated to a general cognitive function in the past, on the other hand, this perspective has been challenged with the rehabilitation of the brain accordingly with its exercise frequency. In other words, a decrease in cognition may occur at a slower rate and damaged neurons may recover or even replaced by proximate neurons (Goh & Park, 2009).

There are a number of age-related changes (Czaja et al., 2013; Vaz-Serra, 2006), in which brain-training games may play a key role. These are:

- Increased difficulty in understanding long and/or understand complex terms.
- Difficulty in solving problems that demand logic abstract analysis and information selection.
- Slowdown in performing psychomotor tasks that involve velocity and short reaction time.
- Loss of memory and decrements in multitasking capacity, selective, and divided attention.
- Difficulties in performing activities that demand spatial orientation, reasoning, numerical, or verbal skills.

These cognitive changes and its implications for designing brain-training/game-based learning are summarized in Table 3.3. As shown in this table, brain-training games can be beneficial to overcome some age-related changes in memory and attention, such as losses in working memories, processing speed, executive functions, and attention (Nouchi et al., 2012).

Although these changes are important, there are also other important aspects to take into account when designing games (Schell & Kaufman, 2016), namely the availability of the game equipment, game instructions, training, social opportunities and inform the older adult players of the possible advantages with the game-playing activity.

Based on these changes mentioned in Table 3.3 and playtesting of brain-training digital games on the market (e.g., Cards, Dominoes, Four in a Line, and Word Search) with nine participants from daycare centers aged between 70 and 89 years old (M = 80; SD = 6.66) during a 10-week session, the following brain-training games were developed: (a) memory game; (b) word search; (c) *Sudoku*; and (d) connect four.

These games were then play-tested by another group of 10 participants from a University of Third Age, aged between 61 and 80 years (M = 69.5; SD = 5.4) and who did not participate previously in the 10-week game playing intervention and design process. They assessed the games in terms of clarity, theme and adequacy of information, game elements and environment, and interaction mechanisms.

(a) *Memory game*

Memory games are the most popular games among older adults, given the importance of memory capacity in the participants' daily lives. For the

Table 3.3 Implications for designing brain-training games/game-based learning based on age-related changes (Adapted from Costa, 2013; Czaja & Sharit, 2012; Veloso & Costa, 2014)

Age-related changes	Implications in game design
Information processing is not as fast as in youth	Lessen cognitive load and divide information into simple words and pictures. Design instructions for different learning paces.
Loss of short-term memory (working memory)	Train players' working memory through the repetition of in-game missions. Avoid large amounts of information and relate new knowledge with previous experiences. Incorporate new learning (neuroplasticity-new neurons have stronger connections). Learning experiences can stimulate episodic memory by acting as third "evocative" places and challenging spatial cognition.
Declines in semantic memory (e.g., historical/cultural facts) and spatial cognition	Connect new learning to real-life events. Familiarize learners with the learning content. Learning objects can activate memories and emotions Spatial memory and navigation can be activated by learning environments.
Declines in procedural memory (know how to perform certain tasks)	Foster procedural memory through the use of mimesis, simulations, and training tutorials. Encourage Learn-It-Yourself and Do-It-Yourself philosophies; Stimulate the learners to build and exchange their experiences by bringing moments of their lives into the learning environment.
Decrements in selective and divided attention	Train visual selective attention skills and decision-making (neuroplasticity-new neurons have stronger connections).
Declines in speech, language comprehension or numerical skills	Encourage reading comprehension and spelling through game narratives and reading materials.

game visuals, a wood texture was used as background to allude players to a traditional context- "Playing cards on table" (Figure 3.7).

In this game, the number of cards that is set on the table is dependent on the current game level (Figure 3.8) and each level shows the kings of the different dynasties of Portugal: The first and second game levels are related with the House of Burgundy; The third and fourth levels refer to the House Aviz; and the fifth and sixth levels cover the Philippine Dynasty and Houze of Braganza, respectively.

When advancing from each game level, players are congratulated by different kings and queens from different dynasties, who send them messages.

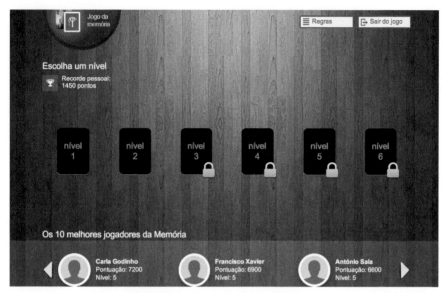

Figure 3.7   Game levels-memory game.

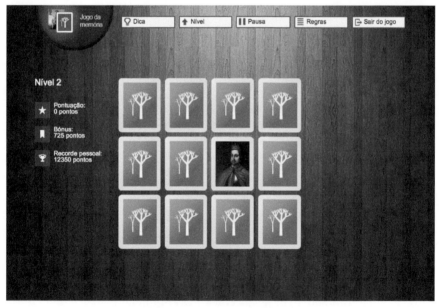

Figure 3.8   Memory game-cards.

During game-playing, most of the participants found that the game was accessible and clear. However, the mechanism of unlocking levels (see Figure 3.8) was not easily understood and, therefore, a caption or tutorial relative to the act of unlocking levels would be needed.

(b) *Word search*

Word search games have a great potential to train the players' attention, spatial memory and word recognition. Indeed, each letter is scanned and examined in detail before checking any pattern and comparing the word detected with the one asked in the word search game (Rayner & Fisher, 1987).

The aim of this game is to circle hidden words in a grid. A list with the words to look for is provided to the player and when a word is found, it is checked from the list. These words may be vertically, horizontally, or diagonally hidden.

For the game visuals, a newspaper pattern was used as background and the themes of the word search game included: names of the Portuguese rivers, Portuguese districts, countries and fruits (themes suggested by the first group of participants) randomly generated (Figure 3.9).

Drag-and-drop interaction was also replaced by clicking on the first and last letter, given that participants experienced some difficulty in performing drag-and-drop.

During game-playing, most of the participants found the game very pleasing and would recommend it to a friend.

(c) *Sudoku*

The Sudoku game is beneficial to improve players' cognition-that is problem-solving, attention, and short-term memory (Ferreira, Owen, Morhan, et al., 2014).

Relative to the game visuals, an ancient paper and bamboo were used as background to create a Japanese puzzle environment (Figure 3.10).

This brain-training digital game provides immediate feedback on the players' answer (marked on red if wrong), with implications on the game points.

During gameplay, players found that the rules were clear and the information presented was adequate. Out of these seven players, five played Sudoku on paper, being more satisfied with the digital game than the paper version because of immediate feedback and the sense of progression through the use of levels.

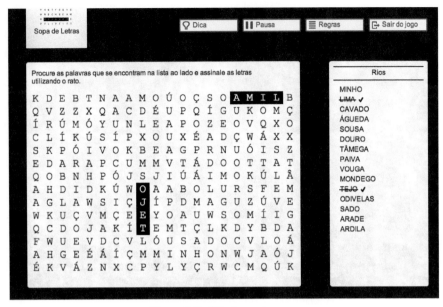

Figure 3.9   Word search game.

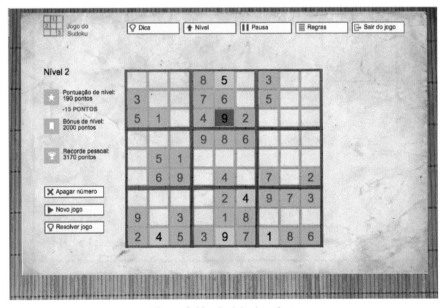

Figure 3.10   Sudoku interface.

(d) *Connect four*

This brain-training connect four promotes social interactions within community members. Single- and multi-player versions were developed, with the option of inviting players to the game.

This game is turn-based and players select one of the two discs (yellow and red). The main goal is to form a line either vertically, horizontally or diagonally of four discs of the same color before the opponent.

In this game (Figure 3.11), players click on the column they want the disc falls.

Special attention had to be given to the indication of whose turn to play was and the option to pause the game.

During the game-playing session, a general lack of familiarity with this game was observed and, hence, only half of the participants were satisfied with the game. For future developments, a mini-tutorial level was suggested.

In view of all these developed brain-training games, the following design recommendations could be drawn:

- Encourage themes of memory and retrospection of the living past, history narrative and share of past experiences.

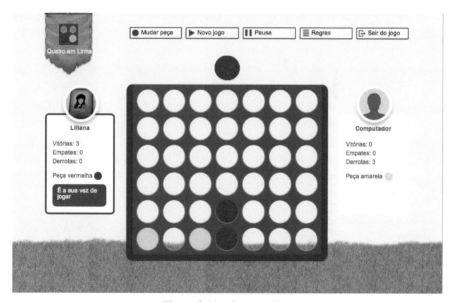

Figure 3.11 Connect four.

- Promote players' familiarity with the game by using game elements that are analogous with the physical environment and context of use.
- Foster social interactions between players (e.g., multiplayer games) and if the game is turn-based, clearly indicate whose turn it is to play.
- Provide game instructions, mini-tutorials, and tips.

Beyond brain-training games, home-based exergames can also be key to contribute to active ageing.

3.4.4 Exergames and Older Adult's Fitness

In the light of the strategy adopted by Nintendo Wii to consider heterogeneous players and widen the gameplay experience to the whole family, there has been an increased interest in the older adult gamer profile (Pearce, 2008) and the adoption of body input- or gesture-based user interfaces-UI.

Exergames or/and Fitness games may defined as the use of games to encourage physical activities, mainly through the use of mimesis, that is the use of the human capacity to imitate others' behaviors (Herrero, 2011) and evolution of game controllers to the virtualization of body movements and simple gestures (Juul, 2012), accessible to different generations of players.

Although exercise-based games have been popularized with such games as *Kinect Sports, Wii Fit*, and *Dance Dance Revolution* on the market and advances in *Kinect[6]* and *Leap Motion[7]* technologies, prior attempts to combine game-playing and physical activity have been made. The following is an account of the many initiatives performed in the exergame market:

- The *Atari Puffer Project[8]* that consisted of a fitness bike digital game for physical activity addressed to sedentary and obese children and adults (Sinclair, Hingston, & Masek, 2007). Although this project from the 1980s was not commercialized, other exergames followed since then.
- *Highcycle[9]* was a game with a stationary bicycle as a controller (Park et al., 2014; Sinclair et al., 2007). The interaction paradigm has evolved to integrate foot pads, and carpets with built-in sensors.

---

[6] Kinect https://developer.microsoft.com/pt-pt/windows/kinect/ (Access Date: Dec 1, 2020).

[7] Ultra Leap https://www.ultraleap.com/product/leap-motion-controller/ (Access Date: Dec 1, 2020).

[8] AGH'S Atari Project Puffer Page, http://www.atarihq.com/othersec/puffer/ (Access Date: Dec 1, 2020).

[9] Autodesk HighCycle http://serious.gameclassification.com/EN/games/45049-HighCycle /index.html(Access date: Dec 3, 2020).

- *Power Pad[10]* and Dance *Dance Revolution (DDR)[11]* pave the way into dance games, being DDR a commercial success.[12]
- *Nintendo Wii* [13] includes the *Wii Controller*, which integrates motion sensors (i.e., accelerometer and gyroscope) to detect the players' actions and position. A series of sport games were then launched, that is tennis, baseball, bowling, golf, and boxing.
- Advances in players' movement tracking and democratization of home-based fitness games, for example, the *Microsoft Kinect* and *Leap Motion*.

In view of home-based exergames or/and fitness, the following recommendations should be considered to be both attractive and effective (Loos, 2017; Sinclair, Hingston, & Masek, 2007):

- Presentation of workout plans.
- Customization of exercises that meet different players' skills, and abilities.
- Biometric feedback-weight management, energy expenditure, health, and exercise competition.
- Aerobic sessions-Warmup, Stimulus, and Cooldown.
- Safety measures, socialization and replay activities.

Considering that exergaming relies heavily on body motion and language and these basic needs are important to Human interaction. Understanding kinesics behaviors and the use of space in game-playing interactions is, therefore, essential given its role in communication and negotiation.

According to Birdwhistell (1952), kinesis is defined as "the movement of the human body in its higher-level activities as a member unit in the cultural context." Poyatos (2002, p. 204) adds the following:

> *"Conscious and unconscious psych muscularly-based body movements and intervening or resulting still positions, either learned or somatogenic, of visual, visual-acoustic and tactile and kinesthetic perception, which, whether isolated or combined with the linguistic and paralinguistic structures and with other somatic and objectual behavioral systems, possess intended or unintended communicative value."*

---

[10]Power Pad, https://nintendo.fandom.com/wiki/Power_Pad (Access date: Dec 3, 2020).

[11]DDR Game, https://www.ddrgame.com (Access date: Dec 3, 2020).

[12]Dance Dance Revolution hits 6.5 million in sales, https://www.gamespot.com/articles/dance-dance-revolution-hits-65-million-in-sales/1100-6084894/ (Access date: Dec 3, 2020).

[13]Wii, https://www.nintendo.pt/Wii/Wii-94559.html (Access date: Dec 3, 2020).

When considering these human body movements either innate (e.g., eye blinking and facial flushing), socio-cultural (e.g., expressions, gestures, and body language), or a combination of both (e.g., laughing, shrugging, and shoulders) (Danesi, 2004) in the context of games, space semiotics and kinesic cues should be considered in game design.

The application of body input- or gesture-based movements in game-playing activities can be, hence, classified into the following (Costa et al., 2018):

- Facial expressions and gaze movements are explored through, for example, the of use of camera to guide the players' journey and recall of what the post-gameplay experience has been.
- Gesture and hand movements are particularly relevant in exergames in order to represent the gameplaying activity (e.g., grab and throw a ball).
- Manual contact indicators with partially controlled actions-for example, indicate actions as sorting, dropping and distinguish blocking areas from interacting ones.
- Posture, in which gestural user interfaces may be also indicators of posture and physical actions are mimetized to perform certain activities.
- Free movement actions: When the players' movement actions are free and they may be driven by missions and supplies' locations.
- Distal and proximal movements are relative to the position and distance of the players in relation to artifacts (e.g., statues, monuments, and historic buildings) or other gamers. A set of Locative and force indicators may also determine the interactions between players and artefacts or surrounding environment.

Collectively, these gesture categories are important in game design and assess the player's interactions with different game scenarios. In fact, these gestures can be of three kinds:

- Semiotic gestures: These gestures result from community meaning attribution and collective experiences. According to McNeil (1992), these gestures can (a) describe an object with semantic content (iconic) or abstract ideas (metaphors); (b) point; and (c) indicate directions (up/down; in/out). An example of a semiotic gesture is the thumbs up to indicate confirmation and approval.
- Ergodic gestures: These gestures have the purpose of manipulating physical objects. This type of gestures depends on the object's state (solid, liquid, and gaseous), change of the object (shape, position, and direction), and whether one or two players' hands are used. An example of this type of gestures is the throwing of a bowling ball.

• Epistemic gestures: These gestures result from tactile experiences with the surrounding environment.

Considering that game-mediated interactions are also characterized by the way players relate in space, another aspect that is crucial in exergame design is proxemics. This latter concept has been introduced by Edward Hall, 1969, who draws our attention to four types of space that differ in terms of distance/proximity: (1) intimate space; (2) personal space; (3) social space; and (4) public space. When covering proxemics in game design, on- and off-screen interactions play a key role.

Figure 3.12 shows the model of proxemics in game-mediated interaction based on the Hall's (1969) spaces, which determine the relational distance within the game context. For these interactions, game designers should take into account the players' positions and devices, size of devices, number of players, types of games, and feedback.

This model of proxemics in game-mediated interactions shown in Figure 3.12 is divided into "hot mediated interactions" (the ones established in the same physical space) and "cool mediated interactions" (the ones established in different physical spaces). From this model, one may suppose that as the distance between players increases (blue and green colors), the player's interactions is more mediated. By contrast, their interaction is more natural as players share the same space (yellow and red colors).

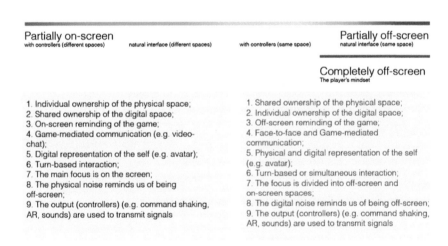

Figure 3.12   Proxemics model in game-mediated interactions.

In "cool mediated interactions," which are increasingly mediated, are often characterized by the following: individual ownership of the physical space; shared ownership of the digital space; on-screen reminding of the game activity; use of game-mediated communication (videochat, instant messaging); digital representation of the self; use of turn-based interactions; and the players' focus on the screen, whereas the physical noise act as reminders of off-screen stimuli.

By contrast, the characteristics that define partially off-screen interactions are the following: Shared ownership of the physical space; individual ownership of the digital space; off-screen reminding of the game activity; face-to-face and game-mediated communications; physical and digital representations of the self; turn-based or simultaneous interactions; and the focus is divided into off-screen and on-screen spaces, whereas the digital noise reminds us of being off-screen.

In both scenarios, the output controllers (e.g., command shaking, augmented reality and sounds) are used to transmit signals.

Based on the aforementioned recommendations for designing home-based exergames, the use of different types of gestures and proxemics, a virtualization of the Portuguese traditional game "Malha"[14] (Figure 3.13) was carried out, addressed the target group of older adults.

For developing this game, a group of five participants were involved in a set of 10 design sessions (M = 80.2; SD = 2.6) by helping to identify the design requirements with the playtest of Kinect Adventures and Sports, whereas other five participants tested the game "Malha" (M = 78.8; SD = 7.2). Most of the participants had 4 years of schooling and gameplay observations enabled to assess the easiness to understand the command options, pause the game, and recognize and perform gestures.

During gameplay testing, there were the following difficulties: (a) Understand some game conventions used in menu navigation (semiotic gestures); (b) perform abduction and adduction exercises, in comparison with flexion and extension exercises; and (c) perform point and wait exercises or other involving static positions and precise movements.

In all, the following exergame design practices can be outlined:

• Explore nostalgic themes and shared experiences.

---

[14]The "Malha" exergame is an adaptation of the physical and traditional Portuguese game "Malha" to a gesture-based interaction paradigm. In this game, players throw a disc toward a pin with the purpose of overthrowing it and leaving a mesh close to the pin.

Figure 3.13 Exergame "Malha" (Terra & Veloso, 2014).

- Provide a game interface in which the players would be familiar with by establishing an analogy with actions that may be performed in daily life.
- Avoid static positions (Nielsen, Storring, Moeslund, et al., 2004) and determine a minimum dimension for each game element (nonprecise movements may be mistaken with involuntary movements-e.g., Parkinson).
- Avoid position and movement combinations that may create some tiredness (Ketcham & Stelmach, 2004; Saffer, 2009).
- Sync game scenarios with the players' conceptual model.

Having presented the role that games can take in rehabilitation, brain training and fitness, the following section will discuss some of the accessibility aspects to take into account in designing these games.

3.5 Accessibility in Games

Accessibility issues have also become of great importance to make games more playable by a wider audience. According to the association of Ablegamers (Barlet & Spohn, 2012), 60% of adults use accessible features on computers even if they do not need them. Nevertheless, there is a cost associated with accessibility and usability. These worries demand time investment and knowledge in both areas. Despite the fact that the return on investment may be low, customer loyalty can be achieved through usability and accessibility evaluation.

The term accessibility describes opening games up to the disabled. Many people with impairments are excluded from computer games because of accessibility problems. Initially, games' accessibility has been regarded as an area of minor importance, although, games had been developed under the slogan "designed for all." Games should be playable by people with and without disabilities (Glinert & York, 2008) because among the users, it may be people who are impaired and people who are unable to use a mouse or keyboard and rely on AT.

The Ablegamers association published an accessibility checklist, which can help adding accessibility for disabled gamers. For example, giving the option of customizing the keyboard, the camera or mouse sensitivity control can help with the mobility of players.

The most successful games are those in which participants can start at a level (Weisman, 1995) and there are some recommendations that can improve video games in terms of accessibility (IGDA, 2004). Kurniawan and Zaphiris (2005) also report some of these applied to other multimedia applications. These recommendations are:

- Provide a high contrast between the font and the background colour;
- Establish a minimum size of 3D objects, lower the density of objects in images and display elements closer to the center.
- Simplify instructions, enhance the social component and create new learning.
- Allow the self-regulation of the levels of difficulty and the ability to move and resize the interface elements.
- Provide remapable keys, customizable configurations and the option of saving the game.
- For first person shooters (fps) and Massive Multiplayer Online Role-Playing Games-MMORPGs, the user has to be able to control the speed, angle and distance of the character in relation to the field of view.
- Display sound and dialogs in the form of text.
- Guide the user through the game by in-game tutorials and provide extra feedback.

Some of these recommendations were considered in the video game Dragon Age. In 2009, it was awarded with the title of 2009 Mainstream Accessible game of the year by presenting full subtitles, multiple alternative controls, diverse color schemes, a "click-to-move" interface, the ability to pause the game and auto-save features (IGDA, 2004).

In the specific case of older adults, there are some other guidelines in the Eldergames project (Gamberini et al., 2006, 2009) to review: (a) Add categorization processes and problem solving in order to maximize the abilities of selective and divided attention as well as short-time memory; (b) predominate collaborative dynamics over competition; (c) integrate a 3D perspective of game elements with the aim of improving visual-spatial orientation, logical reasoning, perception, attention, concentration and strategic ability. Finally, (d) enlarge the visual symbols.

In the light of different types of disabilities that may affect older adults, the following subsections summarize a set of accessibility issues that address visual, auditory, mobility and cognitive abilities. These guidelines are based on the previous work of Bierre and colleagues (2005), Cairns, Power, Barlet, & Haynes, 2019, Ellis and colleagues (2020), Grammenos, Savidis and Stephanidis, 2009, Microsoft, 2018, and Miesenberger and colleagues (2008).

3.5.1 Visual Abilities

Visual impairments are one of the biggest challenges faced with aging (World Health Organization, 2012). Indeed, there is a general difficulty in adapting to darkness and sudden changes in illumination or perceiving low contrast (Fisk, Rogers, Charness, Czaja, & Sharit, 2009). In addition, there is difficulty in focusing and reduction in peripheral vision, field of view and speed processing (Haegerstrom-Portnoy, Schneck, & Brabyn, 1999).

The following guidelines may be, therefore, applied to both enhance the visual experience and overpass some of the visual impairments

- Balance game scenarios in terms of darkness and light, being consistent with the amount of light/darkness in different scenarios.
- Do not rely only on color to highlight and differentiate information and other elements.
- Use a readable font size and contrast between text and background.
- Provide color blind options.
- Establish a minimum size for game elements.
- Set the field of view of the camera to a reasonable view environment as default and give the option to adjust it;
- Include voiceovers for every text, audio descriptions or/and voiced GPS and screen readers.
- Design virtual reality environments that avoid motion sickness.

3.5.2 Auditory Abilities

Auditory impairments often involve peripheral and pure-tone hearing loss beyond a general difficulty in perceiving speech sounds distorted by noise (Helfer & Freyman, 2008).

To enhance the gameplay auditory experience, the following guidelines are suggested:

- Subtitle game cutscenes and speech.
- Provide the option to control and mute game music, speech, and effects.
- Support both text, symbol-based chat and voice for multiplayer games.
- Identify the sources of sound and speakers.
- Provide both stereo/mono options.

3.5.3 Tactile/Motor Abilities

There are also some changes in tactile and motor abilities with the ageing process-that is decrease in sensitivity of touch, vibration, and temperature; loss of kinesthetic sensitivity; difficulties in mobility, and loss of movement precision (Berg, 2016; Fisk et al. , 2009).

The following guidelines are, therefore, suggested to enhance the game-play tactile experience and mobility:

- Do not depend the game challenges on speed/time, avoid "twitch"[15] games.
- Give the possibility to change keyboard settings or/and the option of either interacting with keyboard or cursor or to adjust the sensitivity of game controls.
- Enable portrait and landscape gameplay.
- Offer the possibility to play games, using different types of input-for example, pads, speech commands, gestures, keys/presses, eye-tracking, and avoid simultaneous actions-for example, swipe, click, drag and drop.
- Provide alternatives to body and motion tracking challenges.

3.5.4 Cognitive and Linguistic Abilities

The aging process is often associated to a decline in certain cognitive functions and difficulty in performing tasks that involve reaction time, motor coordination, short-term memory, and abstract or complex reasoning (Czaja

---

[15]Games that test reaction time.

& Sharit, 2012; Van Hooren et al., 2007). In addition, there is general difficulty in numerical skills, speech and language comprehension.

Given these changes, the following guidelines are recommended:

- Provide in-game contextual help and instructions.
- Remind the player of the game goals and controls during the gameplay.
- Give players the option to save the game and facilitate its access through the use of gameplay saves.
- Accompany text information with visuals and voice.
- Highlight key words of game instructions and game goals.
- Offer the possibility to use short words as speech recognition instructions-for example, "yes," "no," and "enter."

In view of these visual, auditory, tactile, and cognitive abilities, the aforementioned considerations are essential to the game industry to make games accessible and ensure better integration of older adults into the game community and a good gameplay experience (Archambault et al., 2008).

3.6 Concluding Remarks

This chapter presented the design considerations to develop games for active ageing and, including game accessibility. The application of these in cognitive and physical rehabilitation is also covered.

When addressing market-oriented products for an ageing population, a general focus on health-related dimensions has been verified, overlooking other pillars of active ageing, that is security and participation in society.

Certain aspects affecting the older adults' interactions with digital devices were also highlighted, namely the access to ICT, perception of the need to use digitally mediated devices, and difficulties given age bio-psycho-social changes. Furthermore, the assumption of older adults' cognitive models is very likely to affect interface design, generating some bias instead of ensuring the users' involvement in the ideation and design process.

Games can constitute "neo-literate" environments and, as such, the following issues should be considered: content and literacy activities, phygital (physical and digital) environments, socio-cultural context, assessment terms and collaborative partnerships.

In fact, games have been applied in key areas essential to health, mobility, intergenerational learning and assistive living. For that, accessibility is essential.

Overall, games may impact human potential and well-being and this chapter covered the way interface design can embody a much humanistic approach and bring a positive playing experience.

References

Abeele, V. V., & De Schutter, B. (2010). Designing intergenerational play via enactive interaction, competition and acceleration. Personal and Ubiquitous Computing, 14(5), 425–433. DOI 10.1007/s00779-009-0262-3

Aghabeigi, B. (2011). Understanding and evaluating cooperative video games [Doctoral dissertation], Communication, Art & Technology: School of Interactive Arts and Technology.

Alben, L. (1996). Quality of experience: defining the criteria for effective interaction design. Interactions, 3(3), 11–15. https://dl.acm.org/doi/10.11 45/235008.235010

Alfadhli, S., & Alsumait, A. (2015, December). Game-based learning guidelines: Designing for Learning and Fun. In 2015 International Conference on Computational Science and Computational Intelligence (CSCI) (pp. 595–600). IEEE. DOI: 10.1109/CSCI.2015.37

Alves, M. L., Mesquita, B. S., Morais, W. S., Leal, J. C., Satler, C. E., & dos Santos Mendes, F. A. (2018). Nintendo Wii^{TM} Versus Xbox Kinect^{TM} for Assisting People With Parkinson's Disease. Perceptual and motor skills, 125(3), 546–565. https://doi.org/10.1177/0031512518769204

ASLERD (2016). Timiosara Declaration Better Learning for a Better World through People Centred Smart Learning Ecosystems, pp. 1–9. Retrieved from http://www.aslerd.org/inevent/events/Aslerd/?s=222 (Access date: Jan 9, 2021)

Alzheimer's Disease International (2013).World Alzheimer Report 2013. Journey of Caring-An Analysis of long-term care for dementia. Retrieved from https://www.alzint.org/u/WorldAlzheimerReport2013.pdf Access date: Dec. 29, 2020)

Archambault D., Gaudy T., Miesenberger K., Natkin S., Ossmann R. (2008) Towards Generalised Accessibility of Computer Games. In: Pan Z., Zhang X., El Rhalibi A., Woo W., Li Y. (eds) Technologies for E-Learning and Digital Entertainment. Edutainment 2008. Lecture Notes in Computer Science, vol 5093. Springer, Berlin, Heidelberg. https://doi.org/10.1007/978-3-540-69736-7-55

Alzheimer's Disease International (2010). World Alzheimer Report 2010-The Global Economic Impact of Dementia. Retrieved from https://www.alz.org/documents/national/worldalzheimerreport2010.pdf (Access date: Dec. 29, 2020)

Aşın, A., Atar, E., Koçyiğit, H., & Tosun, A. (2018). Effects of Kinect-based virtual reality game training on upper extremity motor recovery in chronic stroke. Somatosensory & motor research, 35(1), 25–32. https://doi.org/10.1080/08990220.2018.1444599

Assistive Technology Act of 2004. Pub. L. 108–364. Retrieved from https://www.govinfo.gov/content/pkg/STATUTE-118/pdf/STATUTE-118-Pg1707.pdf#page=31 (Access date: Jan 10, 2020)

Barlet, M.; Spohn, S.D. (2012). *Includification-A Practical Guide Game Accessibility* [Report, Drumgoole and Mason (Eds.)]. Retrieved from https://accessible.games/wp-content/uploads/2018/11/AbleGamers-Includification.pdf (Access date: Dec. 29, 2020)

Balsis, S., Carpenter, B. D., & Storandt, M. (2005). Personality change precedes clinical diagnosis of dementia of the Alzheimer type. The Journals of Gerontology Series B: Psychological Sciences and Social Sciences, 60(2), P98-P101. https://doi.org/10.1093/geronb/60.2.P98

Benavente, A., Rosa, A., Costa, A., Ávila, P. (1996). *A literacia em Portugal: resultados de uma pesquisa extensiva e monográfica.* Lisbon, PT: Fundação Calouste Gulbenkian, Conselho Nacional de Educação.

Ben-Sadoun, G., Manera, V., Alvarez, J., Sacco, G., & Robert, P. (2018). Recommendations for the design of serious games in neurodegenerative diseases. F*rontiers in Aging Neuroscience*, 10, 13, 1–7. https://doi.org/10.3389/fnagi.2018.00013

Benveniste, S., Jouvelot, P., & Péquignot, R. (2010, September). The MINWii Project: Renarcissization of patients suffering from Alzheimer's Disease through video game-based music therapy. In International Conference on Entertainment Computing (pp. 79–90). Springer, Berlin, Heidelberg. https://doi.org/10.1007/978-3-642-15399-08

Berg, J. (2016). Mobility changes during the first years of retirement. *Quality in Ageing and Older Adults*, Vol. 17 No. 2, pp. 131–140. https://doi.org/10.1108/QAOA-11-2015-0052

Best, B. J., & Lebiere, C. (2006). Cognitive agents interacting in real and virtual worlds. *Cognition and multi-agent interaction: From cognitive modeling to social simulation*, New York: Cambridge University Press, pp.186–218.

Bickenbach, J., Cieza, A., Rauch, A., & Stucki, G. (Eds.). (2012). ICF core sets: manual for clinical practice for the ICF research branch, in cooperation with the WHO collaborating centre for the family of international classifications in Germany (DIMDI). Hogrefe Publishing.

Bieńkiewicz, M. M., Brandi, M. L., Hughes, C., Voitl, A., & Hermsdörfer, J. (2015). The complexity of the relationship between neuropsychological deficits and impairment in everyday tasks after stroke. *Brain and Behavior*, 5(10), e00371. https://doi.org/10.1002/brb3.371

Bierre, K., Chetwynd, J., Ellis, B., Hinn, D. M., Ludi, S., & Westin, T. (2005, July). *Game not over: Accessibility issues in video games*. Games Accessibility Special Interest Group, International Game Developers Association, 1–10.

Birdwhistell, R. L. (1952). Introduction to Kinesics. Louisville, USA: University of Louisville.

Breazeal C., Dautenhahn K., Kanda T. (2016) Social Robotics. In: Siciliano B., Khatib O. (eds) Springer Handbook of Robotics. Springer Handbooks. Springer, Cham. https://doi.org/10.1007/978-3-319-32552-172

Bruce, C. S. (2004). Information literacy as a catalyst for educational change. A background paper. In Danaher, Patrick A. *"Lifelong Learning: Whose responsibility and what is your contribution?"*, the 3rd International Lifelong Learning Conference, 13–16 June 2004, Yeppoon, Queensland.

Burke, J. W., McNeill, M. D. J., Charles, D. K., Morrow, P. J., Crosbie, J. H., & McDonough, S. M. (2009). Optimising engagement for stroke rehabilitation using serious games. The Visual Computer, 25(12), 1085–1099. DOI 10.1007/s00371-009-0387-4

Cabrera, M., Malanowski, N. (2009). Ageing Societies, Information and Communication Technologies and Active Ageing, In Cabrera, M., & Malanowski, N. (Eds.). (2009). Information and communication technologies for active ageing: opportunities and challenges for the European Union (Vol. 23). Amestardam, NL: IOS Press, 1–6

Cairns, P., Power, C., Barlet, M., & Haynes, G. (2019). Future design of accessibility in games: A design vocabulary. *International Journal of Human-Computer Studies*, 131, 64–71. https://doi.org/10.1016/j.ijhcs.2019.06.010

Calvo, R. A., & Peters, D. (2014). *Positive Computing: Technology for Wellbeing and Human Potential*. Massachusetts, USA:The MIT Press.

Cassell, J. (2000). Nudge Nudge Wink Wink: Elements of Face-to-Face Conversation for Embodied Conversational Agents. In J. Cassell et al. (Eds.) Embodied Conversational Agents, Cambridge, MA: The MIT Press.

Castells, M. (2001). *The Internet galaxy: Reflections on the Internet, business, and society*. Oxford, UK: Oxford University Press, Inc.

Cecil, R., Parahoo, K., Thompson, K., McCaughan, E., Power, M., & Campbell, Y. (2011). 'The hard work starts now': a glimpse into the lives of carers of community-dwelling stroke survivors. Journal of Clinical Nursing, 20(11–12), 1723–1730. https://doi.org/10.1111/j.1365-2702.2010.03400.x

Centers for Disease Control and Prevention (2020a). About Stroke. Retrieved from https://www.cdc.gov/stroke/about.htm (Access date: Dec. 29, 2020)

Centers for Disease Control and Prevention (2020b). Types of Stroke. Retrieved from https://www.cdc.gov/stroke/typesofstroke.htm (Access date: Dec. 29, 2020)

Chaldogeridis A., Tsiatsos T., Gialaouzidis M., Tsolaki M. (2014) Comparing Data from a Computer Based Intervention Program for Patients with Alzheimer's Disease. In: Shumaker R., Lackey S. (eds) Virtual, Augmented and Mixed Reality. Applications of Virtual and Augmented Reality. VAMR 2014. Lecture Notes in Computer Science, vol 8526. Springer, Cham. https://doi.org/10.1007/978-3-319-07464-124

Chaudhuri, K. R., Odin, P., Antonini, A., & Martinez-Martin, P. (2011). Parkinson's disease: the non-motor issues. *Parkinsonism & related disorders*, 17(10), 717–723. https://doi.org/10.1016/j.parkreldis.2011.02.018

Chaudhuri, K. R., Healy, D. G., & Schapira, A. H. (2006). Non-motor symptoms of Parkinson's disease: diagnosis and management. *The Lancet Neurology*, 5(3), 235–245. https://doi.org/10.1016/S1474-4422 (06)70373-8

Chauveau, L. A., Szilas, N., Luiu, A. L., & Ehrler, F. (2018, May). Dimensions of personalization in a narrative pedagogical simulation for Alzheimer's caregivers. In 2018 *IEEE 6th International Conference on Serious Games and Applications for Health* (SeGAH) (pp. 1–8). IEEE. DOI: 10.1109/SeGAH.2018.8401354

Glanz, K., Rimer, B.K., Viswanath, K. (Eds.) (2008). Health Behavior and Health Education: Theory, Research, and Practice. San Francisco, USA: Wiley.

Cohen, G. D., Firth, K. M., Biddle, S., Lloyd Lewis, M. J., & Simmens, S. (2009). The first therapeutic game specifically designed and evaluated for Alzheimer's disease. *American Journal of Alzheimer's Disease & Other Dementias®*, 23(6), 540–551. https://doi.org/10.1177/15333175 08323570

Connolly, T., Stansfield, M. H., & Hainey, T. (2008, October). Development of a general framework for evaluating games-based learning. In Proceedings of the 2nd European conference on games-based learning (pp. 105–114). Universitat Oberta de Catalunya Barcelona, Spain.

Cook, B., & Twidle, P. (2016, October). Increasing Awareness of Alzheimer's Disease through a Mobile Game. In 2016 International Conference on Interactive Technologies and Games (ITAG) (pp. 55–60). IEEE. https://doi.org/10.1109/itag.2016.16

Cooper, A., Reimann, R., Cronin, D., & Noessel, C. (2014). About Face: The Essentials of Interaction Design, 4th Edition. Indianapolis, USA: Wiley.

Costa, L.V., Veloso, A.I., Mealha, Ó. (2018). Citizen's interactions in "Smart Game-Playing Environments", In D. King (Eds.) 19th International Conference on Intelligence Games and Simulation, GAME-ON 2018, pp. 121–126.

Costa, L. (2013). Networked video games for older adults [Master's Thesis] University of Aveiro, Aveiro, Portugal. Retrieved from http://ria.ua.pt/handle/10773/11326 (Access date: Dec. 28th, 2020)

Costa, L. V., Veloso, A. I., Loizou, M., & Arnab, S. (2018, May). Breaking barriers to game-based learning for active ageing and healthy lifestyles A qualitative interview study with experts in the field. In 2018 IEEE 6th International Conference on Serious Games and Applications for Health (SeGAH) (pp. 1–8). IEEE. DOI: 10.1109/SeGAH.2018.8401320

Costa, L., & Veloso, A. (2016). Being (Grand) Players: Review of Digital Games and their Potential to Enhance Intergenerational Interactions. *Journal of Intergenerational Relationships*, 14(1). https://doi.org/10.1080/15350770.2016.1138273

Csikzentmihalyi, M. (2008).The Psychology of Optimal Experience. New York, USA: Harper Perennial Modern Classics.

Czaja, S. J., Sharit, J., Lee, C. C., Nair, S. N., Hernández, M. A., Arana, N., & Fu, S. H. (2013). Factors influencing use of an e-health website in a community sample of older adults. *Journal of the American Medical Informatics Association*, 20(2), 277–284. doi: 10.1136/amiajnl-2012-000876

Czaja, S. J., & Sharit, J. (2012). *Designing training and instructional programs for older adults*. Boca Raton, FL: CRC Press.

Damasio, A. (2003). *Looking for Spinoza: Joy,Sorrow, and the Feeling Brain. Sorrow, and the Feeling Brain*. New York, USA: Random House.

Danesi, M. (2004). *Messages, signs, and meanings: A Basic Textbook in Semiotics and Communication*. Vol. 1, Toronto, Canada: Canadian Scholars' Press.

De la Hera, T., Loos, E.F., Simons, M, & Blom, J. (2017). Benefits and Factors Influencing the Design of Intergenerational Digital Games: A Systematic Literature Review, *Societies* 7(18). https://doi.org/10.3390/soc7030018

De Melo Cerqueira, T. M., de Moura, J. A., de Lira, J. O., Leal, J. C., D'Amelio, M., & do Santos Mendes, F. A. (2020). Cognitive and motor effects of Kinect-based games training in people with and without Parkinson disease: A preliminary study. *Physiotherapy Research International,* 25(1), e1807. https://doi.org/10.1002/pri.1807

Desurvire, H., Caplan, M., & Toth, J. A. (2004, April). Using heuristics to evaluate the playability of games. In CHI'04 *extended abstracts on Human factors in computing systems* (pp. 1509–1512). https://doi.org/10.1145/985921.986102

Dewhurst, E., Novakova, B., & Reuber, M. (2015). A prospective service evaluation of acceptance and commitment therapy for patients with refractory epilepsy. *Epilepsy & Behavior,* 46, 234–241. https://doi.org/10.1016/j.yebeh.2015.01.010

de Oliveira Santos, L. G. N., Ishitani, L., & Nobre, C. N. (2013). Casual mobile games for the elderly: a usability study. SBC-Proceedings of SBGames, São Paulo, October 16th and 18th, 2013.

Dias, J., Veloso, A. I., & Ribeiro, T. (2019). "A Priest in the Air": Developing a motion-based and immersive digital game for stroke rehabilitation, *2019 14th Iberian Conference on Information Systems and Technologies (CISTI)*, 19–22 June 2019, Coimbra, Portugal. doi:10.23919/cisti.2019.8760748

Dias, S., Ioakeimidis, I., Dimitropoulos, K., Grammatikopoulou, A., Grammalidis, N., Diniz, J. A., ... & Hadjileontiadis, L. J. (2020, June). Innovative interventions for Parkinson's disease patients using iPrognosis games: an evaluation analysis by medical experts. In Proceedings of the *13th ACM International Conference on PErvasive Technologies Related to Assistive Environments* (pp. 1–8). https://doi.org/10.1145/3389189.3397974

Earman, J. (1986). A primer on determinism (Vol.32). NL: Springer Netherlands

Ellis, B., Ford-Williams, G., Graham, L., Grammenos, D., Hamilton, I., Lee, E, Manion, J., Westin, T. (2020). Game accessibility guidelines [Website]. Retrieved from http://gameaccessibilityguidelines.com/why-and-how/ (Access date: December 29, 2020)

Encalada, P., Medina, J., Manzano, S., Pallo, J. P., Chicaiza, D., Gordón, C., ... & Andaluz, D. F. (2019, September). Virtual Therapy System in a Multisensory Environment for Patients with Alzheimer's. In: Bi Y., Bhatia R., Kapoor S. (eds) *Intelligent Systems and Applications. IntelliSys 2019. Advances in Intelligent Systems and Computing,* vol 1038. Springer, Cham. https://doi.org/10.1007/978-3-030-29513-457

European Commission (2019). Policies for Ageing Well with Information and Communication Technologies (ICT). Retrieved from https://ec.europa.eu /digital-single-market/en/ageing-well (Access date: Jan. 6th, 2021)

European Parkinson Disease Association (2020). What is Parkinson's? Retrieved from https://www.epda.eu.com/about-parkinsons/what-is-p arkinsons/ (Access date: December 29, 2020)

European Union (2017). European Urban Mobility-Policy Context. Retrieved from https://ec.europa.eu/transport/sites/transport/files/2017-sustainable-urban-mobility-policy-context.pdf (Access date: January 9, 2021)

Evensen, E. A. (2009). Making it fun: Uncovering a Design research model for Educational BOARD Game Design (Masterdissertation), The Ohio State University. Retrieved from http://rave.ohiolink.edu/etdc/view?acc-num=osu1247862315 (Access date: January 9, 2021)

Federoff, M. A. (2002). Heuristics and usability guidelines for the creation and evaluation of fun in video games [Doctoral dissertation], Indiana University, Indiana. Retrieved from http://citeseerx.ist.psu.edu/viewdoc/download?doi=10.1.1.89.8294&rep=rep1&type=pdf Access date: January 9, 2021)

Felsted, K. F.,& Wright, S. D. (2014). *Toward Post Ageing: Technology in an Ageing Society (Vol. 1). Switzerland: Springer International Publishing.*

Ferreira, N., Owen, A., Mohan, A., Corbett, A.,& Ballard, C. (2015). Associations between cognitively stimulating leisure activities, cognitive function and age-related cognitive decline. International journal of geriatric psychiatry, 30(4), 422–430. https://doi.org/10.1002/gps.4155

Ferreira, S., Torres, A., Mealha, Ó., & Veloso, A.I. (2014). Training Effects on Older People in Information and Communication Technologies Considering Psychosocial Variables. Educational Gerontology, 41(7), 482–493. Doi https://doi.org/10.1080/03601277.2014.994351

Fisk, A. D., Rogers, W. A., Charness, N., Czaja, S. J., Sharit, J. (2009). Designing for Older Adults: Principles and Creative Human Factors Approaches, 2nd ed. Human Factors & Aging Series, Boca Raton: CRC Press.

Francillette, Y., Boucher, E., Bouchard, B., Bouchard, K., & Gaboury, S. (2021). Serious games for people with mental disorders: State of the art of practices to maintain engagement and accessibility. Entertainment Computing, 37, 100396. doi:10.1016/j.entcom.2020.100396

Frank, C., Pari, G., & Rossiter, J. P. (2006). Approach to diagnosis of Parkinson disease. *Canadian family physician*, 52(7), 862–868.

Gamberini, L., Raya, M. A., Barresi, G., Fabregat, M., Ibanez, F., & Prontu, L. (2006). Cognition, technology and games for the elderly: An introduction to ELDERGAMES Project. *PsychNology Journal*, 4(3), 285–308.

Gamberini, L., Martino, F., Seraglia, B., Spagnolli, A., Fabregat, M., Ibanez, F., ... & Andrés, J. M. (2009, May). Eldergames project: An innovative mixed reality table-top solution to preserve cognitive functions in elderly people. In *2009 2nd conference on human system interactions* (pp. 164–169). IEEE. DOI: 10.1109/HSI.2009.5090973

García-Betances, R. I., Arredondo Waldmeyer, M. T., Fico, G., & Cabrera-Umpiérrez, M. F. (2015). A succinct overview of virtual reality technology use in Alzheimer's disease. Frontiers in aging neuroscience, 7, 80. https://doi.org/10.3389/fnagi.2015.00080

Gavriushenko, M., Karilainen, L., & Kankaanranta, M. (2015). Adaptive systems as enablers of feedback in English language learning game-based environments. 2015 *IEEE Frontiers in Education Conference* (FIE). doi:10.1109/fie.2015.7344107

Geyh, S., Cieza, A., Schouten, J., Dickson, H., Frommelt, P., Omar, Z., ... & Stucki, G. (2004). ICF Core Sets for stroke. *Journal of rehabilitation medicine*, 36(0), 135–141. DOI: 10.1080/16501960410016776

Gilbert, N., & Conte, R. (1995). Artificial societies. Taylor & Francis. DOI: 10.4324/9780203993699

Glanz, K., Rimer, B. K., & Viswanath, K. (2008). Health behavior and health education: theory, research, and practice. New Jersey, USA: John Wiley & Sons.

Glinert, E. P., & York, B. W. (2008). Computers and people with disabilities. A*CM Transactions on Accessible Computing* (TACCESS), 1(2), 1–7. https://doi.org/10.1145/1408760.1408761

Goetz, C. G., Tilley, B. C., Shaftman, S. R., Stebbins, G. T., Fahn, S., Martinez-Martin, P., ... & LaPelle, N. (2008). Movement Disorder Society-sponsored revision of the Unified Parkinson's Disease Rating Scale (MDS-UPDRS): scale presentation and clinimetric testing results. Movement disorders: official journal of the Movement Disorder Society, 23(15), 2129–2170.

Goh, J. O., & Park, D. C. (2009). Neuroplasticity and cognitive aging: the scaffolding theory of aging and cognition, *Restorative Neurology and Neuroscience*, 27(5), 391- 403. DOI: 10.3233/RNN-2009-0493

Grammenos, D., Savidis, A., & Stephanidis, C. (2009). Designing universally accessible games. *Computers in Entertainment* (CIE), 7(1), 1–29. https://doi.org/10.1145/1486508.1486516

Gavriushenko, M., Karilainen, L., & Kankaanranta, M. (2015, October). Adaptive systems as enablers of feedback in English language learning game-based environments. In 2015 IEEE Frontiers in Education Conference (FIE) (pp. 1–8). IEEE.

Griffiths, M., Kuss, D. J., & de Gortari, A. B. O. (2013). Videogames as therapy: A review of the medical and psychological literature. In *Handbook of Research on ICTs and Management Systems for Improving Efficiency in Healthcare and Social Care*, ch.3, pp. 43–68. IGI Global. doi: 10.4018/978-1-4666-3990-4.ch003

Haegerstrom-Portnoy, G., Schneck, M. E., & Brabyn, J. A. (1999). Seeing into old age: vision function beyond acuity. *Optometry and vision science*, 76(3), 141–158.

Hall, E. T.(1969). *The hidden dimension*. New York, USA: Anchor Books.

Harwood, J. (2007). *Understanding Communication and Aging: Developing Knowledge and Awareness*. University of Arizon, USA: SAGE Publications.

Helfer, K. S., & Freyman, R. L. (2008). Aging and speech-on-speech masking. *Ear and hearing*, 29(1), 87.

Herrero, L. (2011). Homo Imitans: The Art of Social Infection: Viral Change in Action. UK: MeetingMinds

Holling, C.S. (2001) Understanding the complexity of economic, ecological, and social systems. *Ecosystems*, 4(5), p. 390–405. doi: 10.1007/s10021-001-0101-5

Ibrahim, R., & Jaafar, A. (2009, August). Educational games (EG) design framework: Combination of game design, pedagogy and content modeling. In 2009 international conference on electrical engineering and informatics (Vol. 1, pp. 293–298). *IEEE*. DOI: 10.1109/ICEEI.2009.5254771

IGDA-International Game Developers Association (2004).*Accessibility in Games: Motivations and Approaches* [Report]. IGDA Game Accessibility SIG:

International Longevity Centre Brazil (2015). Active Ageing: A Policy Framework in Response to the Longevity Revolution. Rio de Janeiro: International Longevity Centre Brazil, 2015, p.1–117.

ISO 9241-11:1998, Ergonomic requirements for office work with visual display terminals (VDTs)-Part 11: Guidance on usability (1998) by ISSO

Jameson, E., Trevena, J., & Swain, N. (2011). Electronic gaming as pain distraction. Pain Research and Management, 16(1), 27-32.

Jordan, P. (2000). *Designing Pleasurable Products: An Introduction to the New Human Factors*. London, UK: Taylor and Francis.

Juul, J. (2012). A Casual Revolution: Reinventing Video Games and Their Players. Massachusets, USA: MIT Press.

Kalisch, H. R., Coughlin, D. R., Ballard, S. M., & Lamson, A. (2013). Old age is a part of living: Student reflections on intergenerational service-learning. *Gerontology & Geriatrics Education*, 34(1), 99–113. https://doi.org/10.1080/02701960.2012.753440

Kalliopuska, M. (1994). Relations of retired people and their grandchildren. *Psychological reports*, 75(3), 1083–1088. https://doi.org/10.2466/pr0.19 94.75.3.1083

Kempler, D. (1995). Language changes in dementia of the Alzheimer type. *Dementia and communication*, 98–114.

Kenner, C., Ruby, M., Jessel, J., Gregory, E., and Arju, T. (2007). Intergenerational learning between children and grandparents in East London. *Journal of Early Childhood Research*, 5(3), 219–243. https://doi.org/10.1177/1476718X07080471

Ketcham, C. J., & Stelmach, G. E. (2004). Movement control in the older adult. *National Research Council. 2004. Technology for adaptive aging*, Washington, DC: The National Academies Press, pp. 64–92.

Khan, A. M., & Lawo, M. (2016). Wearable Recognition System for Emotional States Using Physiological Devices. In *Proceedings of the 8th International Conference on eHealth, Telemedicine, and Social Medicine* (eTELEMED), 131–137

Kyeong, S., Kang, H., Kyeong, S., & Kim, D. H. (2019). Differences in brain areas affecting language function after stroke. *Stroke*, 50(10), 2956–2959.

Klijn, C. J., & Hankey, G. J. (2003). Management of acute ischaemic stroke: new guidelines from the American Stroke Association and European Stroke Initiative. The Lancet Neurology, 2(11), 698–701. https://doi.org/10.1016/S1474-4422(03)00558-1

Koohi, N. (2014). Hearing loss in stroke. Auditory event-related potentials to words: Implications for audiologists. Hearing Health Matters. Retrieved from https://hearinghealthmatters.org/pathways/2014/hearing-loss-stroke/ (Access date: Dec. 29, 2020)

Korhonen, H., & Koivisto, E. M. (2006, September). Playability heuristics for mobile games. In Proceedings of the 8th conference on Human-computer interaction with mobile devices and services (pp. 9–16). https://doi.org/10.1145/1152215.1152218

Kornhaber, A., & Woodward, K. L. (1981) . *Grandparents, grandchildren: The vital connection*. New York, USA: Routledge.

Kroma, A., & Lachman, R. (2018, October). Alzheimer's Eyes Challenge: The Gamification of Empathy Machines. In Proceedings of the 2018 Annual Symposium on Computer-Human Interaction in Play Companion Extended Abstracts (pp. 329–336). https://doi.org/10.1145/3270316.3270320

Kurniawan, S., & Zaphiris, P. (2005). Research-Derived Web Design Guidelines for Older People, *Proceedings of the ACM SIGACCESS Conference on Computers and Accessibility, ASSETS 2005*, Baltimore, MD, USA, October 9–12, pp. 1–8. DOI: 10.1145/1090785.1090810

Kurtis, M. M., Rodriguez-Blazquez, C., Martinez-Martin, P., & ELEP Group. (2013). Relationship between sleep disorders and other non-motor symptoms in Parkinson's disease. *Parkinsonism & related disorders*, 19(12), 1152–1155. https://doi.org/10.1016/j.parkreldis.2013.07.026

Lancaster, C., Koychev, I., Blane, J., Chinner, A., Chatham, C., Taylor, K., & Hinds, C. (2020). Gallery Game: Smartphone-based assessment of long-term memory in adults at risk of Alzheimer's disease. Journal of Clinical and Experimental Neuropsychology, 42(4), 329–343. https://doi.org/10.1080/13803395.2020.1714551

Langhorne, P., Bernhardt, J., & Kwakkel, G. (2011). Stroke rehabilitation. *The Lancet*, 377(9778), 1693–1702. Doi: 10.1016/S0140-6736 (11)60325-5

Law E.LC., Kickmeier-Rust M.D., Albert D., Holzinger A. (2008) Challenges in the Development and Evaluation of Immersive Digital Educational Games. In: Holzinger A. (eds) HCI and Usability for Education and Work. USAB 2008. Lecture Notes in Computer Science, vol 5298. Springer, Berlin, Heidelberg. https://doi.org/10.1007/978-3-540-89350-92

Law, E. L. C., & Sun, X. (2012). Evaluating user experience of adaptive digital educational games with Activity Theory. International Journal of Human-Computer Studies, 70(7), 478–497. https://doi.org/10.1016/j.ijhcs.2012.01.007

Lefebvre, H. (1996). Writing on Cities [Translated and Edited by Eleonore Kofman and Elizabeth Lebas], Oxford, UK: Blackwell Publishers Ltd

Leung, L., & Lee, P. S. (2005). Multiple determinants of life quality: The roles of Internet activities, use of new media, social support, and leisure activities. *Telematics and Informatics*, 22(3), 161–180. https://doi.org/10.1016/j.tele.2004.04.003

Lewis, G. N., Woods, C., Rosie, J. A., & Mcpherson, K. M. (2011). Virtual reality games for rehabilitation of people with stroke: perspectives from the users. Disability and Rehabilitation: *Assistive Technology*, 6(5), 453–463.doi: 10.3109/17483107.2011.574310

Lin, J., He, J., Shu, B., & Jia, J. (2020). Multi-sensory feedback therapy combined with task-oriented training on the hemiparetic upper limb in chronic stroke: study protocol for a pilot randomized controlled trial. *ResearchSquare*, DOI: https://doi.org/10.21203/rs.3.rs-27503/v1

Lodha, N., Naik, S. K., Coombes, S. A., & Cauraugh, J. H. (2010). Force control and degree of motor impairments in chronic stroke. *Clinical Neurophysiology*, 121(11), 1952–1961. https://doi.org/10.1016/j.clinph.2010.04.005

Loetscher, T., Potter, K.J., Wong, D., das Nair, R. (2019). *Cognitive rehabilitation for attention deficits following stroke*. Cochrane Database of Systematic Reviews 2019, Issue 11. Art. No.: CD002842. DOI: 10.1002/14651858.CD002842.pub3. Accessed 29 December 2020.

Loos, E. F. (2017). Exergaming: Meaningful Play for Older Adults? In J. Zhou, & G. Salvendy (Eds.), In J. Zhou, & G. Salvendy (Eds.), Human Aspects of IT for the Aged Population. Healthy and Active Aging, Second International Conference, ITAP 2017, Held as Part of HCI International 2017 Vancouver, Canada, July 9–14, 2017, Proceedings, Part II Applications, Services and Contexts (pp. 254–265). Switzerland: Springer International Publishing.

Marin, J. G., Lawrence, E., Navarro, K. F., & Sax, C. (2011). Heuristic evaluation for interactive games within elderly users. In Proceedings of the 3rd International Conference on eHealth, Telemedicine, and Social Medicine (eTELEMED'11) (pp. 130–133).

Marinelli, E. C., & Rogers, W. A. (2014, September). Identifying potential usability challenges for xbox 360 kinect exergames for older adults. In Proceedings of the Human Factors and Ergonomics Society Annual Meeting (Vol. 58, No. 1, pp. 1247–1251). Sage CA: Los Angeles, CA: SAGE Publications. https://doi.org/10.1177/1541931214581260

Martyr, A., & Clare, L. (2012). Executive function and activities of daily living in Alzheimer's disease: a correlational meta-analysis. *Dementia and geriatric cognitive disorders*, 33(2–3), 189–203. https://doi.org/10.1159/000338233

Mayasari, A., Pedell, S., & Barnes, C. (2016, November). "Out of sight, out of mind", investigating affective intergenerational communication over distance. In Proceedings of the 28th Australian Conference on Computer-Human Interaction (pp. 282–291). https://doi.org/10.1145/3010915.3010937

Mc Namara, K., Alzubaidi, H., & Jackson, J. K. (2019). Cardiovascular disease as a leading cause of death: how are pharmacists getting involved?. Integrated pharmacy research & practice, 8, 1–11. https://doi.org/10.2147/IPRP.S133088

McNeill, D. (1992).*Hand and mind: What gestures reveal about thought.* Illinois, USA: University of Chicago Press.

Michie, S., Van Stralen, M. M., & West, R. (2011). The behaviour change wheel: a new method for characterising and designing behaviour change interventions. *Implementation science*, 6(1), 1–12. doi:10.1186/1748-5908-6-42

Microsoft (2018). Making Video Games Accessible: Business Justifications and Design Considerations. Retrieved from https://docs.microsoft.com/pt-pt/windows/win32/dxtecharts/accessibility-best-practices?redirectedfrom=MSDN (Access date: October 29, 2020)

Miesenberger K., Ossmann R., Archambault D., Searle G., Holzinger A. (2008) More Than Just a Game: Accessibility in Computer Games. In: Holzinger A. (eds) HCI and Usability for Education and Work. USAB 2008. *Lecture Notes in Computer Science, vol 5298*. Springer, Berlin, Heidelberg. https://doi.org/10.1007/978-3-540-89350-918

Miller, A. P., Navar, A. M., Roubin, G. S., & Oparil, S. (2016). Cardiovascular care for older adults: hypertension and stroke in the older adult. *Journal of geriatric cardiology: JGC*, 13(5), 373–379. doi: 10.11909/j.issn.1671-5411.2016.05.001

Morris, R. G., & Kopelman, M. D. (1986). The memory deficits in Alzheimer-type dementia: A review. *The Quarterly Journal of Experimental Psychology*, 38(4), 575–602. https://doi.org/10.1080/14640748608401615

Muscio, C., Tiraboschi, P., Guerra, U. P., Defanti, C. A., & Frisoni, G. B. (2015). Clinical trial design of serious gaming in mild cognitive impairment. *Frontiers in aging neuroscience*, 7, 26. https://doi.org/10.3389/fnagi.2015.00026

Nacke, L., Drachen, A., Kuikkaniemi, K., Niesenhaus, J., Korhonen, H. J., Hoogen, W. M., ... & De Kort, Y. A. (2009). Playability and player experience research. In Proceedings of digra 2009: Breaking new ground: Innovation in games, play, practice and theory. DiGRA.

National Institute on Aging (2019). Alzheimer's Disease Fact Sheet. Retrieved from https://www.nia.nih.gov/health/alzheimers-disease-fact-sheet (Access date: January 4, 2021)

Navarro, K. F., Lawrence, E., Marin, J. G., & Sax, C. (2011). A dynamic and customisable layered serious game design framework for improving the physical and mental health of the aged and the infirm. In Conference on eHealth, Telemedicine, and Social Medicine. IARIA Conference. ISBN: 978-1-61208-119-9

Nef, T., Chesham, A., Schütz, N., Botros, A. A., Vanbellingen, T., Burgunder, J. M., ... & Urwyler, P. (2020). Development and Evaluation of Maze-Like Puzzle Games to Assess Cognitive and Motor Function in Aging and Neurodegenerative Diseases. *Frontiers in aging neuroscience*, 12, 87. https://doi.org/10.3389/fnagi.2020.00087

Ng, Y. Y., Khong, C. W., & Thwaites, H. (2012). A review of affective design towards video games. Procedia-Social and Behavioral Sciences, 51, 687–691. https://doi.org/10.1016/j.sbspro.2012.08.225

Nielsen, M., Störring, M., Moeslund, T. B., & Granum, E. (2004). A Procedure for Developing Intuitive and Ergonomic Gesture Interfaces for HCI. Lecture Notes in Computer Science, 409–420. doi:10.1007/978-3-540-24598-838

Norman, D. A. (2005). Human-centered design considered harmful. Interactions, 12(4), 14–19. doi:10.1145/1070960.1070976

Norman, D. A. (2004). *Emotional Design: Why We Love (or Hate) Everyday Things*. New York, USA: Basic Books.

Nouchi, R., Taki, Y., Takeuchi, H., Hashizume, H., Akitsuki, Y., Shigemune, Y., & Yomogida, Y. (2012). Brain training game improves executive functions and processing speed in the elderly: A randomized controlled trial. PLoS ONE, 7(1), e29676. doi:10.1371/journal.pone.0029676 PMID:22253758

OECD (2015). Ageing in Cities-Policy Highlights. Retrieved from oecd.org/cfe/regionaldevelopment/Policy-Brief-Ageing-in-Cities.pdf (Access date: January 9, 2021)

Oña, E. D., Jardón, A., Cuesta-Gómez, A., Sánchez-Herrera-Baeza, P., Cano-de-la-Cuerda, R., & Balaguer, C. (2020). Validity of a Fully-Immersive VR-Based Version of the Box and Blocks Test for Upper Limb Function Assessment in Parkinson's Disease. Sensors, 20(10), 2773. https://doi.org/10.3390/s20102773

Olson, D. R., & Torrance, N. (Eds.). (2009). *The Cambridge handbook of literacy*. Cambridge, UK:Cambridge University Press.

Paavilainen, J. (2010, May). Critical review on video game evaluation heuristics: social games perspective. In Proceedings of the International Academic Conference on the Future of Game Design and Technology (pp. 56–65). https://doi.org/10.1145/1920778.1920787

Paavilainen, J., Kultima, A., Kuittinen, J., Mäyrä, F., Saarenpää, H., & Niemelä, J. (2009). GameSpace: Methods and Evaluation for Casual Mobile Multiplayer Games, University of Tampere. Retrieved from https://trepo.tuni.fi/handle/10024/65773 (Access date: Jan 10, 2021)

Papaloukas, S., Patriarcheas, K., & Xenos, M. (2009, September). Usability assessment heuristics in new genre videogames. In 2009 13th Panhellenic Conference on Informatics (pp. 202–206). IEEE. DOI: 10.1109/PCI.2009.14

Pachoulakis, I., Xilourgos, N., Papadopoulos, N., & Analyti, A. (2016). A Kinect-based physiotherapy and assessment platform for Parkinson's disease patients. *Journal of medical engineering*, 2016. https://doi.org/10.1155/2016/9413642

Paletta, L., Pszeida, M., Dini, A., Russegger, S., Schuessler, S., Jos, A., ... & Fellner, M. (2020, June). MIRA-A Gaze-based Serious Game for Continuous Estimation of Alzheimer's Mental State. In *ACM Symposium on Eye Tracking Research and Applications* (pp. 1–3). https://doi.org/10.1145/3379157.3391989

Paletta L., Pszeida M., Panagl M. (2019) Towards Playful Monitoring of Executive Functions: Deficits in Inhibition Control as Indicator for Cognitive Impairment in First Stages of Alzheimer. In: Lightner N. (eds) Advances in Human Factors and Ergonomics in Healthcare and Medical Devices. AHFE 2018. Advances in Intelligent Systems and Computing, vol 779. Springer, Cham. https://doi.org/10.1007/978-3-319-94373-2-12

Papaloukas, S., Stoli, C., Patriarcheas, K., & Xenos, M. (2010, September). A survey on how online games usability affects the retention of information. In 2010 14th Panhellenic Conference on Informatics (pp. 209–213). IEEE, 10–12 Sept. 2010. DOI: 10.1109/PCI.2010.11

Parnell, M. J., Berthouze, N., & Brumby, D. (2009). *Playing with scales: Creating a measurement scale to assess the experience of video games.* University College London, London, UK.

Park, T., Lee, U., MacKenzie, S., Moon, M., Hwang, I., & Song, J. (2014, April). Human factors of speed-based exergame controllers. In *Proceedings of the SIGCHI Conference on Human Factors in Computing Systems* (pp. 1865–1874). https://doi.org/10.1145/2556288.2557091

Pearce, C. (2008). The truth about baby boomer gamers: A study of over-forty computer game players. *Games and Culture*, 3(2), 142–174. https://doi.org/10.1177/1555412008314132

Pedersen, P. M., Vinter, K., & Olsen, T. S. (2004). Aphasia after stroke: type, severity and prognosis. *Cerebrovascular diseases*, 17(1), 35–43. https://doi.org/10.1159/000073896

Perry, R. J., & Hodges, J. R. (1999). Attention and executive deficits in Alzheimer's disease: A critical review. *Brain*, 122(3), 383–404. https://doi.org/10.1093/brain/122.3.383

Picard, R. W., & Picard, R. (1997). *Affective computing* (Vol. 252). Cambridge, UK: MIT press.

Pilcher, J. (1994). Mannheim's sociology of generations: an undervalued legacy. *British Journal of Sociology*, 481–495. https://doi.org/10.2307/591659

Pinelle, D., Wong, N., & Stach, T. (2008, April). Heuristic evaluation for games: usability principles for video game design. In Proceedings of the *SIGCHI Conference on Human Factors in Computing Systems* (pp. 1453–1462). https://doi.org/10.1145/1357054.1357282

Pinto, S., Ozsancak, C., Tripoliti, E., Thobois, S., Limousin-Dowsey, P., & Auzou, P. (2004). Treatments for dysarthria in Parkinson's disease. The Lancet Neurology, 3(9), 547–556. https://doi.org/10.1016/S1474-4422(04)00854-3

Poyatos, F. (2002). *Nonverbal Communication Across Disciplines: Paralanguage, kinesics, silence, personal and environmental interaction*. Vol. 2. Amestardam, Netherlands: John Benjamins Publishing Company.

Postman, N. (2011). *Technopoly: The surrender of culture to technology*. New York, USA: Vintage.

Rayner, K., & Fisher, D. L. (1987). Letter processing during eye fixations in visual search. *Perception & Psychophysics*, 42(1), 87–100. https://doi.org/10.3758/BF03211517

Ribeiro, T., Veloso, A. I., & Costa, R. (2016, December). Conceptualization of PhysioFun Game: A low-cost videogame for home-based stroke rehabilitation. In *2016 1st International Conference on Technology and Innovation in Sports, Health and Wellbeing (TISHW)* (pp. 1–8). IEEE. DOI: 10.1109/TISHW.2016.7847787

Rings, S., Steinicke, F., Picker, T., & Prasuhn, C. (2020, March). Memory Journalist: Creating Virtual Reality Exergames for the Treatment of Older Adults with Dementia. In 2020 IEEE Conference on Virtual Reality and 3D User Interfaces Abstracts and Workshops (VRW) (pp. 687–688). IEEE. DOI: 10.1109/VRW50115.2020.00194

Robert, P., König, A., Amieva, H., Andrieu, S., Bremond, F., Bullock, R., ... & Manera, V. (2014). Recommendations for the use of Serious Games in people with Alzheimer's Disease, related disorders and frailty. *Frontiers in aging neuroscience*, 6, 54, 1–13. https://doi.org/10.3389/fnagi.2014.00054

Rong, X., Dahal, S., Luo, Z. Y., Zhou, K., Yao, S. Y., & Zhou, Z. K. (2019). Functional outcomes after total joint arthroplasty are related to the severity of Parkinson's disease: a mid-term follow-up. Journal of orthopaedic surgery and research, 14(1), 396. https://doi.org/10.1186/s13018-019-1447-8

Rowe, F., Brand, D., Jackson, C. A., Price, A., Walker, L., Harrison, S., ... & Howard, C. (2009). Visual impairment following stroke: do stroke patients require vision assessment?. *Age and Ageing*, 38(2), 188–193. https://doi.org/10.1093/ageing/afn230

Sáenz-de-Urturi, Z., García Zapirain, B., & Méndez Zorrilla, A. (2015). Elderly user experience to improve a Kinect-based game playability. *Behaviour & Information Technology*, 34(11), 1040–1051. https://doi.org/10.1080/0144929X.2015.1077889

Sánchez, J. L. G., Vela, F. L. G., Simarro, F. M., & Padilla-Zea, N. (2012). Playability: analysing user experience in video games. *Behaviour & Information Technology*, 31(10), 1033–1054. https://doi.org/10.1080/0144929X.2012.710648

Santos, S. C., & Knijnik, J. D. (2009). Motivos de adesão à prática de atividade física na vida adulta intermediária. Revista Mackenzie de Educação Física E Esporte, 5(1), 23-24.

Saffer, D. (2009). Designing gestural interfaces: touchscreens and interactive devices. Sebastopol, CA: O'Reilly Media, Inc.

Schell, R., & Kaufman, D. (2016). Cognitive Benefits of Digital Games for Older Adults-Strategies for Increasing Participation. Proceedings of the 8th International Conference on Computer Supported Education. doi:10.5220/0005878501370141

Schmidt, M., Paul, S. S., Canning, C. G., Song, J., Smith, S., Love, R., & Allen, N. E. (2020). The accuracy of self-report logbooks of adherence to prescribed home-based exercise in Parkinson's disease. *Disability and Rehabilitation*, 1–8. https://doi.org/10.1080/09638288.2020.1800106

Schrag, A., Hovris, A., Morley, D., Quinn, N., & Jahanshahi, M. (2006). Caregiver-burden in Parkinson's disease is closely associated with psychiatric symptoms, falls, and disability. Parkinsonism & related disorders, 12(1), 35–41. https://doi.org/10.1016/j.parkreldis.2005.06.011

Schuller, T. (2010). Building a Strategic Framework for Lifelong Learning: Insights from "Learning through Life". Adult Learner: The Irish Journal of Adult and Community Education, 105–118.

Seifert, A. (2020). The Digital Exclusion of Older Adults during the COVID-19 Pandemic. Journal of Gerontological Social Work, 1–3. doi:10.1080/01634372.2020.1764687

Siegel, C., & Dorner, T. E. (2017). Information technologies for active and assisted living-Influences to the quality of life of an ageing society. International journal of medical informatics, 100, 32–45. https://doi.org/10.1016/j.ijmedinf.2017.01.012

Sinclair, J., Hingston, P., & Masek, M. (2007, December). Considerations for the design of exergames. In Proceedings of the 5th international conference on Computer graphics and interactive techniques in Australia and Southeast Asia (pp. 289–295). https://doi.org/10.1145/1321261.1321313

Sixsmith, A., and G. Gutman. (2013). "Introductions." In Technologies for Active Aging, 1–5. Boston, MA: Springer. doi:10.1007/978-1-4419-8348-0.

Smith, M.R., & Marx, L. (1994). Does technology drive history? The dilemma of technological determinism. Cambridge, Massachusets: MIT Press.

Spaccavento, S., Marinelli, C. V., Nardulli, R., Macchitella, L., Bivona, U., Piccardi, L., ... & Angelelli, P. (2019). Attention deficits in stroke patients: The role of lesion characteristics, time from stroke, and concomitant neuropsychological deficits. *Behavioural neurology*, 2019. https://doi.org/10.1155/2019/7835710

Szilas, N., Andkjær, K. I., Kemp, L., Ricci, A., Dadema, T., Nap, H. H., & Ehrler, F. (2020). Designing a Senior Friendly Interface for a Personalized 3D Narrative Simulation. International Conference on the Foundations of Digital Games. doi:10.1145/3402942.3403023

Terra, I. & Veloso, A. I., (2014) O "jogo da malha"-videojogo com paradigma de interação gestual adaptado aos seniores. Revista Prisma.com, 23(5), ISSN 1646–3153.

Thwaites, T., Davis, L., & Mules, W. (2002). Introducing cultural and media studies: a semiotic approach. New York, USA: Palgrave.

Torrente, J., Freire, M., Moreno-Ger, P., & Fernández-Manjón, B. (2015). Evaluation of semi-automatically generated accessible interfaces for educational games. *Computers & Education*, 83, 103–117. https://doi.org/10.1016/j.compedu.2015.01.002

Torres, A.C. (2011). Cognitive Effects of Video Games on Old People. International Journal on Disability and Human Development, 10 (1), 55–58. doi: 10.1515/ijdhd.2011.003

Truelsen, T., Ekman, M., & Boysen, G. (2005). Cost of stroke in Europe. European journal of neurology, 12, 78–84. https://doi.org/10.1111/j.1468-1331.2005.01199.x

Uğur, F., & Sertel, M. (2020). The Effect of Virtual Reality Applications on Balance and Gait Speed in Individuals With Alzheimer Dementia: A Pilot Study. *Topics in Geriatric Rehabilitation*, 36(4), 221-229. https://doi.org/10.1097/TGR.0000000000000285

UNESCO (2011). Creating and sustaining literate environments. UNESCO Bangkok Asia and Pacific Regional Bureau for Education, Bangkok: Thailand. Retrieved from http://unesdoc.unesco.org/images/0021/002146/2146 53E.pdf (Access date: Nov 8th, 2017)

UNESCO. (2005). Aspects of literacy assessment. Topics and issues from the UNESCO expert meeting. Paris. Retrieved from http://unesdoc.unesco.org/images/0014/001401/140125eo.pdf (Access date: March8th, 2021)

Vallejo, V., Wyss, P., Rampa, L., Mitache, A. V., Müri, R. M., Mosimann, U. P., & Nef, T. (2017). Evaluation of a novel Serious Game based assessment tool for patients with Alzheimer's disease. PLoS One, 12(5), e0175999. https://doi.org/10.1371/journal.pone.0175999

Vale Costa, L, Veloso, A. I., Loizou, M., Arnab, S., Tomlins, R., & Sukumar, A. (2018). "What a Mobility-Limited World": Design Requirements of an Age-Friendly Playable City. Preprints 2018, 2018100536 (doi: 10.20944/preprints201810.0536.v1).

van Balkom, T.D., Berendse, H.W., van der Werf, Y.D. et al. COGTIPS: a double-blind randomized active controlled trial protocol to study the effect of home-based, online cognitive training on cognition and brain networks in Parkinson's disease. BMC Neuro 119, 179 (2019). https://doi.org/10.1186/s12883-019-1403-6

Van Hooren, S. A. H., Valentijn, A. M., Bosma, H., Ponds, R. W. H. M., Van Boxtel, M. P. J., & Jolles, J. (2007). Cognitive functioning in healthy older adults aged 64–81: a cohort study into the effects of age, sex, and education. *Aging, Neuropsychology, and Cognition*, 14(1), 40–54. https://doi.org/10.1080/138255890969483

Vaz-Serra, A. (2006). *O que significa envelhecer?* In H. Firmino (Ed.), Psicogeriatria (pp. 21–33). Coimbra, PT: Edições Almedina.

Veloso & Costa(2016). Heuristics for designing digital games in assistive environments: Applying the guidelines to an ageing society. 2016 1st International Conference on Technology and Innovation in Sports, Health and Wellbeing (TISHW). doi:10.1109/tishw.2016.7847789

Veloso, A., & Costa, L. (2015, June). Social network games in an ageing society: Co-designing online games with adults aged 50 and over. In 2015 10th Iberian conference on information systems and technologies (CISTI) (pp. 1–6). IEEE. DOI: 10.1109/CISTI.2015.7170613

Veloso, A. I., & Costa, L. (2014). Jogos na comunidade miOne. In A.I. Veloso (Ed.), *SEDUCE Utilização da comunicação e da informação em ecologias web pelo cidadão sénior.* Porto, PT: Ed.Afrontamento

Veloso, A. I., Costa, L., & Ribeiro, T. (2016). Jogos digitais na promoção da saúde: Desafios e tendências. *Revista da FAEEBA-Educação e Contemporaneidade*, 25(46). Doi: 10.21879/faeeba2358-0194.2016.v25.n46.p%p

Vines, J., Pritchard, G., Wright, P., Olivier, P., & Brittain, K. (2015). An Age-Old Problem. ACM Transactions on Computer-Human Interaction, 22(1), 1–27. doi: 10.1145/2696867

Voida, A., & Greenberg, S. (2012). Console gaming across generations: Exploring intergenerational interactions in collocated console gaming. *Universal Access in the Information Society*, 11(1), 45-56. https://doi.org/10.1007/s10209-011-0232-1

Wærstad, M., & Omholt, K. A. (2013). Exercise Games for Elderly People: Identifying important aspects, specifying system requirements and designing a concept (Master's thesis,) Institutt for telematikk, NTNU . Trondheim-Norweigan University of Science and Technology.

Walker, A. (2002). A strategy for active ageing. *International Social Security Review*, 55(1), 121–139. doi: 10.1111/1468-246X.00118

Watkins, C. L., Leathley, M. J., Gregson, J. M., Moore, A. P., Smith, T. L., & Sharma, A. K. (2002). Prevalence of spasticity post stroke. *Clinical rehabilitation*, 16(5), 515-522. DOI: 10.1191/0269215502cr512oa

Watts, P.M. (1982). Nicolaus Cusanus: A Fifteenth-Century Vision of Man. Studies in the History of Christian Tradition, vol. 30. Brill, Leiden

Webber, S., & Johnston, B. (2000). Conceptions of information literacy: new perspectives and implications. *Journal of information science*, 26 (6), 381–397. doi: 10.1177/016555150002600602

Weisman, S. (1995). Computer games for the frail elderly. *Computers in Human Services*, 11(1-2), 229–234.

Wendel, S. (2020). Designing for behavior change: Applying psychology and behavioral economics. 2nds edition. Sebastopol, CA: O'Reilly Media.

Wiederhold, M. D., & Wiederhold, B. K. (2007). Virtual reality and interactive simulation for pain distraction. Pain Medicine, 8 (3), S182-S188. https://doi.org/10.1111/j.1526-4637.2007.00381.x

WHO (2002). Active Ageing: A Policy Framework. Second UN World Assembly on Ageing, Madrid, Spain. Retrieved from https://extranet.w ho.int/agefriendlyworld/wp-content/uploads/2014/06/WHO-Active-Agei ng-Framework.pdf(August1st,2021)

World Health Organization (2020). Dementia-Key facts. Retrieved 04 January 2020 from https://www.who.int/news-room/fact-sheets/detail /dementia

Whyte, J., & Marlow, B. (1999). Beliefs and attitudes of older adults toward voluntary use of the internet: an exploratory investigation. Proc. OzCHI, Wagga Wagga, Australia.

Wright, K. (2000). Computer-mediated social support, older adults, and coping. *Journal of communication*, 50(3), 100–118. https://doi.org/10.1 111/j.1460-2466.2000.tb02855.x

Yuan, R. Y., Chen, S. C., Peng, C. W., Lin, Y. N., Chang, Y. T., & Lai, C. H. (2020). Effects of interactive video-game-based exercise on balance in older adults with mild-to-moderate Parkinson's disease. *Journal of Neuro-Engineering and Rehabilitation*, 17(1), 1–10. https://doi.org/10.1186/s129 84-020-00725-y

Zhang, F., Kaufman, D.A. (2016). Review of intergenerational play for facilitating interactions and learning, *Gerontechnology*, 14, 127–138. https://doi.org/https://doi.org/10.4017/gt.2016.14.3.013.00

Zvacek, S. M. (1991). Effective affective design. *TechTrends*, 36(1), 40–43. doi: https://doi.org/10.1007/BF02761286

Zheng, J., Chen, X., & Yu, P. (2017). Game-based interventions and their impact on dementia: a narrative review. *Australasian Psychiatry*, 25(6), 562–565. https://doi.org/10.1177/1039856217726686

4

Gamification, Senior Tourism, and the Wellness Market

This chapter addresses the use of gamification in senior tourism and the wellness market. It begins by introducing the concept of gamification and its interrelatedness with active learning. Then, the right to leisure and advancements in Senior Tourism are discussed. Finally, gamification in the tourism sector and trends in wellness-oriented markets are presented.

4.1 Defining Gamification

Gamification can broadly be defined as the process of applying game elements and techniques to contexts that go beyond the entertainment purpose (e.g., health, education, security, and tourism) and encourage changes in the player's behaviors and routines (Deterding et al., 2011; Sailer, Hense, Mayr, & Mandl, 2017). As Elias, Rajan, McArthur, and Dasco, (2013, P. 7) put it: Gamification is "the use of game design techniques and mechanics to engage audiences and improve behavior-related outcomes."

It is worth noting that the definition of gamification (game+modification) has not been static over time. From its foundation with the Nick Pelling's enterprise Conundra to the Jesse Schell's presentation on "When Games Invade Real Life and Gamify Work" (Schell, 2010), the definition has been constantly changing to blur the gap between behavior-changes and engagement (Werbach & Hunter, 2012).

The interrelatedness among the concepts of "game thinking," "game elements," and their "context," which characterizes gamification and a gamified strategy (Kapp, 2012; Werbach & Hunter, 2012; Zichermann & Cunningham, 2011) will be covered in the next sections.

4.1.1 Game Thinking

Game thinking refers to a mindset shift to solve problems and the capacity to turn activities fun, adhering to engagement loops, social conflicts *(player-versus-environment, player-versus-player challenges),* and level progression (Kapp, 2012; Werbach & Hunter, 2012).

Game thinking is essential to develop games, gamification, and playful design. Although there has been little discussion about these concepts, there are slight differences that should be highlighted.

According to Deterding et al., (2011), gamification differs from serious games in terms of completeness. Whereas serious games embed game elements and a game design strategy into a nongame context (e.g., goals, challenges, rewards, and rules), these are defragmented in gamification. The playful design is relative to a play activity and interaction (e.g., Google doodles).

In our perspective, the differences in such terms rely mostly on the input or context in which a tedious or boring task becomes fun (i.e., activity, inanimate object, or simulation). Figure 4.1 illustrates the differences between gamification, playful design, and game design.

As shown in Figure 4.1, these processes (i.e., gamification, playful design, and game design) are related to the transformation of a tedious/boring task into fun and transference of game thinking into real-world environments and vice versa. In specific, these are characterized by the following:

- *Gamification* is relative to the transformation of an activity that involves tedious/boring tasks into fun routines. For that, game thinking combined with game elements can lead to sustainable changes in behaviors by providing immediate feedback toward actions.
- *Playful design* is relative to the application of game elements and game thinking to inanimate objects and branding. Whereas one of the ultimate goals of gamification is behavior change, the purpose of playful design is merely aesthetics.
- *Game design* is relative to the representations and simulations of the world and that representation is key to the player's engagement.

When applying game thinking, one should consider the Mechanics—Dynamics—Aesthetics (MDA) framework (Hunicke, LeBlanc & Zubek, 2004). These three dimensions are summarized below:

- *Mechanics* refers to the code, algorithms, and rules that form the game world.

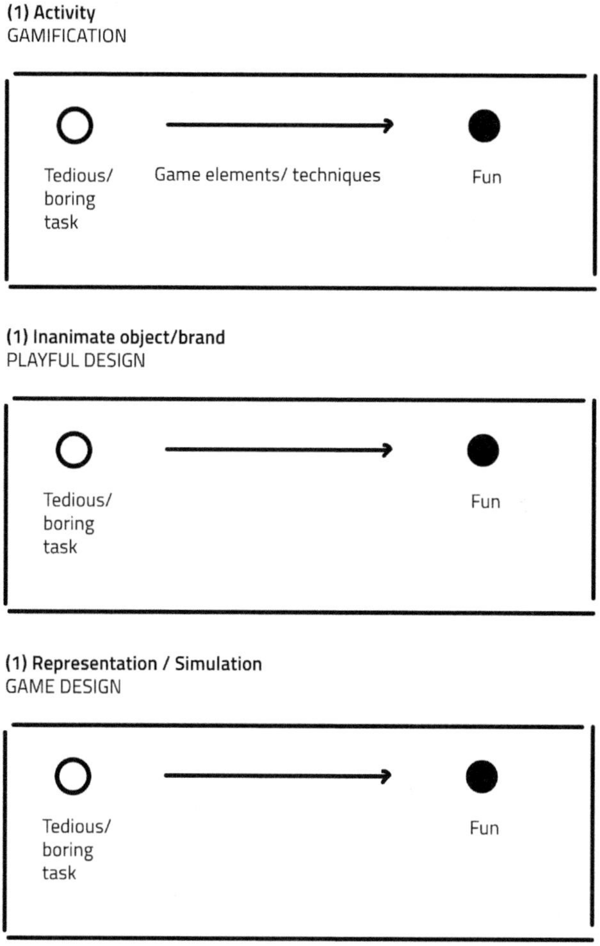

(1) Activity
GAMIFICATION

Tedious/boring task   Game elements/ techniques   Fun

(1) Inanimate object/brand
PLAYFUL DESIGN

Tedious/boring task   Fun

(1) Representation / Simulation
GAME DESIGN

Tedious/boring task   Fun

Figure 4.1 Gamification, playful design and game design.

- *Dynamics* is based on game mechanics, being related to the game procedures and players' interactions.
- *Aesthetics* is relative to the player's emotions during gameplay and, hence, they can be designed through the exploration of sensations, fantasy, narrative, challenges, fellowship, discovery, expression, and submission.

The MDA model was then (re)adopted by Werbach and Hunter (2012, p. 78), who proposed The Game Element Hierarchy or the Dynamics—Mechanics—Components (DMC) model.

- Examples of game dynamics include constraints, emotions, narrative, progression, and relationships.
- Game mechanics include the following elements: challenges, chance, competition, cooperation, feedback, resource acquisition, rewards, transactions, turns, and the win states.
- Some of the game components are achievement, avatars, badges, boss fights, collections, combat, content unlocking, gifting, leaderboards, levels, points, quests, social graphs, teams, and virtual goods.

4.1.2 Game Elements

Beyond game thinking, game elements are also essential to gamify the system. Some of them have been previously mentioned in the DMC model. Based on Malone's work on "What makes things fun to learn?" (Malone, 1980) and the compilation of game elements proposed by Kapp, (2012), Zichermann & Cunningham, (2011), Zichermann and Linder, (2013), and Werbach & Hunter, (2012), Figure 4.2 shows the interrelationship between different game elements within the dimensions of fantasy, curiosity, challenge, and feedback.

The author Malone (1980) suggests three dimensions that engage people in games and the learning process. These dimensions are a challenge, fantasy, and curiosity. However, a fourth dimension that is essential in this framework should be added. That dimension is feedback. A brief description of each dimension is described below:

- *Challenges* imply a sense of mastery and progression in uncertain goals and some randomness.
- *Fantasy* relies on schemas and representations from previous experiences.
- *Curiosity* refers to the optimal level of complexity of the learning environment, being novel, and surprising.
- *Feedback* is relative to certain pieces of immediate information about the players' actions and game performance.

In addition, the combination of these dimensions can activate such neurotransmitters as an internal reward (ephinephirne, nopinephirne, and dopamine) and empathizing capacity (oxytocin; McGonigal, 2011).

Regarding the game elements associated to each dimension, these include:

- *Social feedback and reinforcement:* Pattern recognition, organization tasks, and social feedback are very likely to stimulate the players' brain.

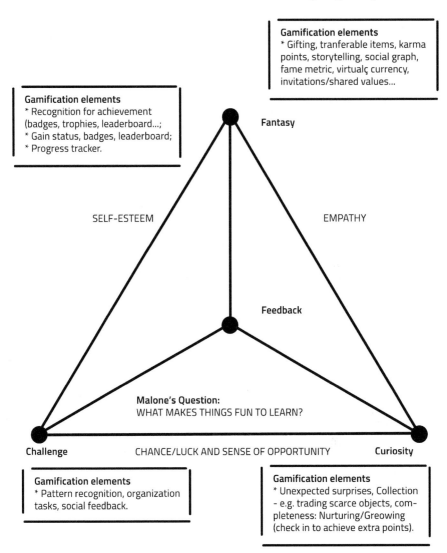

Gamification elements
* Gifting, tranferable items, karma points, storytelling, social graph, fame metric, virtualç currency, invitations/shared values...

Gamification elements
* Recognition for achievement (badges, trophies, leaderboard...;
* Gain status, badges, leaderboard;
* Progress tracker.

Fantasy

SELF-ESTEEM

EMPATHY

Feedback

Malone's Question:
WHAT MAKES THINGS FUN TO LEARN?

Challenge

CHANCE/LUCK AND SENSE OF OPPORTUNITY

Curiosity

Gamification elements
* Pattern recognition, organization tasks, social feedback.

Gamification elements
* Unexpected surprises, Collection - e.g. trading scarce objects, completeness: Nurturing/Greowing (check in to achieve extra points).

Figure 4.2 Gamification mechanics and learning.

Feedback is essential to track the players' progress in the game. Match-three or dinner dash games are some of the games that are based on pattern recognition and organization tasks.

- *Gifts and points:* Unexpected surprises and events can trigger curiosity. Object collection (exploration, scarce objects...), karma, skill, experience, reputation, and redeemable points are some examples. The exchange of these objects and gifting can also establish relationships between players.

- *Storyline:* Storytelling and role-playing are part of empathic design. For that, avatars are usually used as representations of the player.
- *Social graphs and fame's metrics:* Social graphs represent the player's social networks and fame's metrics are often used to order the players based on their visibility—for example, awareness, appeal, aspiration, breakthrough, influence, trendsetter, and trust (Chalanyova & Mikulas, 2017).
- *Achievement, badges, collectibles, and rewards:* Achievements are associated with missions and established goals. These are often represented with badges and other types of collectibles and rewards (e.g., inventory, prizes, and trophies).
- *Challenges, quests, competition, and boss fights:* Player-versus-player and player-versus-environment challenges must be in line with the game purpose. These challenges may both involve competition and collaboration and when associating them with rewards, quests are generated. Later game levels are characterized by increased difficulty, that is boss fights.
- *Rules:* Players' behaviors are determined by game rules.
- *Onboarding, levels, and leaderboards:* Onboarding is the process to induce players to the gameplay. During the game activity, there are three types of game levels associated with the player's progress: missions, difficulty levels, and experience levels. Leaderboards order the players accordingly with their performance and progress in the game.
- *Resources acquisition, virtual goods, content unlocking, and customization:* During gameplay, items can be bought or conquered and transitioned. These items and virtual goods can still have the following attributes: physical location, function, and scarcity. Content unlocking is relative to its availability depending on the players' progress. The game content and items can also be customized, given their properties (e.g., size, color, and shape).

(a) *Nonentertainment contexts*

As mentioned in the gamification definition, nonentertainment contexts play an essential role in the application of game thinking and game elements. Some examples of these common contexts include health, learning and training, news, science and research, financing, and advertising (Deterding et al., 2011).

Having defined what is meant by gamification, the next sections focus on such use in the tourism sector.

4.2 The Senior Tourist and the Right to Leisure

Senior tourism has been attracting a lot of interest within the scientific community (e.g., Alén, Domínguez, & Losada, 2012; Cleaver, Muller, Ruys, & Wei, 1999; Hunter–Jones and Blackburn, 2007; Lieux, Weaver, & McCleary, 1994; Shoemaker, 1989) and tourism sector over the past years. In times of pandemics, the tourism sector has been negatively affected and attention has been directed to "slow" and rural domestic tourism.

In 2015, 48.8% of the 65+ inhabitants from the European Union adhered to tourism activities and this market has been largely driven by the increasing ageing society (World Health Organisation, 2002). Indeed, this segment differs from younger tourists in terms of tourism activities and frequency.

In terms of the distribution of senior tourists out of the total number of European tourists during the year, data from the Eurostat, 2018 reveal that whereas most of the tourists go on vacations during high season (July and August), senior citizens distribute yearly their holiday trips. The older target group also tends to stay away longer in a journey (Lieux, Weaver, & McCleary, 1994) and return to a destination rather than visiting new places (Shoemaker, 1989).

Moreover, cultural tourism that facilitates the access to new knowledge and provides innovative historical and cultural experiences tend to be key to this target group (Patterson, 2006).

A number of studies on the Senior/Silver Tourist (e.g., Cleaver, Muller, Ruys, & Wei, 1999, p. 7; Lieux et al., 1994, p. 725; Sellick, 2004, p. 61; Shoemaker, 1989; Tiago Tiago, de Almeida Couto, Tiago & Faria, 2016, p. 19; You & O'Leary, 1999, p. 27) have proposed a set of tourist profile categories based on their motivations to travel. These are summarized below:

- *Nostalgics, "passive visitors," & "attachment seekers"*: Memorable experiences and a connection between past and present drive these experiences. Closing ties (e.g., relatives and friends) may also be an additional motive to (re) connect with these places (e.g., family's roots, relatives or friend's places) by sharing communal thoughts, feelings, and a sense of belongingness.
- *Friendlies, generational social sharers, "family travelers" and "connection and transcendence seekers"*: These travelers are moved by social

connections and, thus, holiday packages facilitate the sense of togetherness among people or/and create a social bond with grandchildren. Group-oriented interests and values are taken into account in these packages, fostering community awareness and individual's involvement.

- *"Excited learners," "novelty seekers," culture "enthusiastic go-getters," and discovers:* Knowledge seeking, cultural heritage, curiosity, and learning guides the tourist to place visiting and world discovery.
- *Escapists, "active resters" and "indulgent relaxers":* A sense of escapism of daily routines, avoidance of overthinking and boredom, and relaxation are pointed out as the primary reason for traveling.
- *Thinkers, "self-esteem builders" and "spirit and solace seekers":* Introspection and well-being motivate these travelers. Holiday activities may include mindfulness programs and nature-based initiatives, raising one's self-esteem.
- *Status-seekers:* These travelers are motivated by the need to seek a certain status and gain others' respect based on the novelty and number of places visited in some sort of competition with friends and colleagues.
- *Explorers, "physical actives," "active enthusiasts," "livewires," and "vacationers":* These travelers are interested in physical, adventurous, and exciting activities, usually preferring warm weather. Sports and physical activities may motivate these travelers. Spa and wellness experiences make part of this package.
- *"Reluctant travelers," "homebodies," and "older set":* These travelers like trips of shorter duration and spend time with relatives. Health treatments may be additional motives for these travelers.

It is worth noting, however, that there is not a unique profile for defining each traveler but a mix of motives and profiles depending on such subjective variables as a personality trait, mood, context, season, or/and social networks.

Leisure is a Human Right that has been recognized in 1948 with the United Nations Universal Declaration of Human Rights (Karev & Doron, 2017; United Nations General Assembly, 1948). Although leisure has been recognized as a right, these have not had visibility at the international scale (Kaplan, 1986; Karev & Doron, 2017).

Considering that the concept of active ageing relies on the activity theory in that there is the functionalist approach that ageing successfully is associated with the maintenance of active life and social interactions, the right to leisure should also be respected and safeguarded (Karev & Doron, 2017).

Figure 4.3 shows the activity theory model applied to active ageing through the mediation of an artifact. As can be observed in this figure,

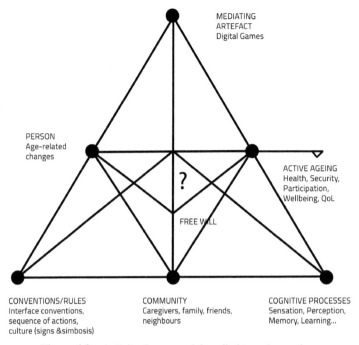

Figure 4.3 Activity theory model applied to active ageing.

a certain individual has an aim (e.g., perceive better health, sense of security and participation in society) and the whole community, conventions used and cognitive processes (sensation and perception) influence the path to achieve this aim. Furthermore, a mediating artifact may affect and facilitate the individual's meaning to the experience of achieving a certain goal.

There is also a risk of these activities being predetermined and influenced by society or cultural values, without free will. Hence, caution must be applied when preparing package activities to the older target group and also respect the right to leisure and recreation. This subchapter has described the profile of senior tourists, highlighting the importance of the right to leisure. The next section will discuss the use of gamification in a tourism setting, providing some examples of its application.

4.3 Gamifying the Tourism Experience

Over the past months, the tourism market has been negatively affected by the COVID-19 crisis and there has been some urgency in re-stimulating confidence in tourism offers and wellness-oriented markets. Game elements and thinking strategies within the tourism sector play a crucial role by simulating travel experiences or/and support decision-making and adherence to a sustainable "slow tourism" attitude (Fullagar, Markwell, & Wilson, 2012).

Gamification for "slow tourism" refers to the use of challenges and other game elements in unexplored places to increase sustainable consumption and length of stays to augment the experience and interaction within the environment (Fullagar, Markwell, & Wilson, 2012). Indeed, there has been an increasing interest in the tourist experience (Bulencea and Egger, 2015) and perceived well-being, addressing intergenerational leisure needs and motivations (Fullagar, Markwell, & Wilson, 2012).

The use of digitally mediated tools in family traveling has also been of interest, prolonging the sense of the place tourists are visiting. Digital media may also augment the tourists' historical and cultural experiences, reinforcing a sense of attachment with the visited place.

Although the use of games in learning and changes in behavior have been widely covered (e.g., Elias, Rajan, McArthur, & Dacso, 2013; Kapp, 2012; Werbach & Hunter, 2012), the reason for applying game elements and thinking into such a taken-for-granted internal-rewarding activity as traveling has been quite overlook. Bulcenea and Egger (2015) claim that its application in a tourism setting can generate memorable experiences and as such, the examples provided in this subchapter aim to foster sustainable tourism, strengthening the connection with local communities and cultural heritage.

The following are some examples of games and gamification apps developed by our students within the context of senior tourism-that is a location-based game for senior tourism, a gamified app for senior cyclotourism, and a newsgame about health tourism and epidemics addressed to older adults.

4.3.1 A Location-based Game for Senior Tourism

A location-based game for senior tourism entitled footour was developed to guide visitors to the Portuguese city Aveiro, using points of interest (POIs) and thematic routes (Figure 4.4). Themes and narratives are attributed to certain routes containing different points of interest that guide the characters, minigames (Figure 4.5), and place storytelling.

A set of requirements for the location-based game were established based on user's tests on an interactive prototype and app benchmarking:

• Representation of POIs in a map.
• Location and nearby important local contacts: police, hospital, and taxi.

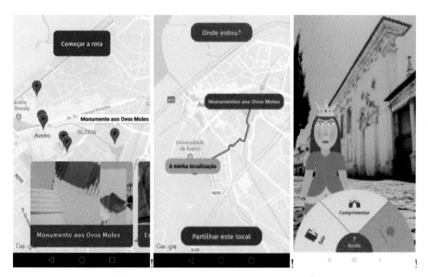

Figure 4.4   Location-based game footour[1].

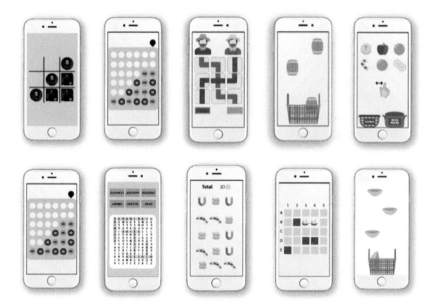

Figure 4.5   Footour mini-games (Veloso et al., 2020).

---

[1]Retrieved https://play.google.com/store/apps/details?id=com.footour.footour&hl=en_US (Access date: Sep. 9, 2020)

- Complementary colors, font's readability, and contrast.
- Routes with historical facts and POIs containing information regarding that spot.
- Route with multiple POIs near each other, given the mobility limitations of the target audience.

Further information about the development of the location game and user testing can be further consulted in the paper *"Footour: Designing and Developing a Location-Based Game for Senior Tourism in the miOne Community"* (Veloso, Carvalho, Sampaio, Ribeiro & Costa, 2020).

Results indicated that when developing a location-based game for senior tourism, the following functionalities would affect the whole experience: Associate historical information to POI; share resources, status, and social activities; and interlink past/present photographs and other content to the visited places and routes.

4.3.2 A Gamified App for Senior Cyclotourism

Jizo is a gamified app to foster senior cyclotourism co-designed with a group of adult learners at a Portuguese University of Third Age (7 focus group participants) and (re) designed with the inputs from a group of 31 cyclists' interviewees).

The co-design process (Muller & Druin, 2003) involved such techniques as PICTIVE (i.e., applying a paper mock-up to simulate the prototype functionalities and interaction with the graphical user interfaces; Muller, 1993), scenario building (i.e., scenarios that simulate the context of the use of the digital media; Carrol, 1999), and collage (i.e., use of pieces of materials to represent a phenomenon or idea, following a pattern; Butler-Kisber and Poldma, 2010; Figure 4.6).

After running this co-design process, the Jizo app was developed accordingly with the following functionalities and elements that were pointed out by the target group (Figure 4.7):

- *Pre-experience:* route recommendations, weather forecast, information about cycling equipment, and bicycle rental shops.
- *In loco:* route directions, performance monitoring, points, achievements, and rewards.
- *Post-experience:* motivation quotes and feedback, unlock resources, badges earnings, rankings, and leaderboards.

Figure 4.6   Jizo co-design process (Ortet et al., 2019).

Figure 4.7   Jizo prototype (Ortet et al., 2019).

Further information about gamification testing can be further consulted in the papers *"Jizo: A Gamified Digital App for Senior Cyclo-Tourism in the miOne Community"* (Ortet et al. 2019) or the master's thesis *Gamification and Senior Cyclo-tourism: Designing an App for the miOne Community* (Ortet, 2019).

4.3.3 A Newsgame about Health Tourism and Epidemics Addressed to Older Adults

Adventour is a newsgame that aims to create awareness to the health risks associated with traveling. The target group of this game is the adult learners at the Universities of the Third Age, in which the game can play an important role by simulating impactful and multidecisional travel experiences on health diseases.

Newsgames are a game type that incorporates news events as game content or journalistic practices and storytelling into a game-based approach (Foxman, 2015), usually simulating a testimonial experience — "sense of being there."

The gamer plays the role of Matilde (Figure 4.8), who accompanies the player in an interactive narrative with several places to discover, challenges to overcome, specific diseases to learn more about and avoid while choosing impactful health decisions and affecting the course of the narrative. During

Figure 4.8   Adventour game description.

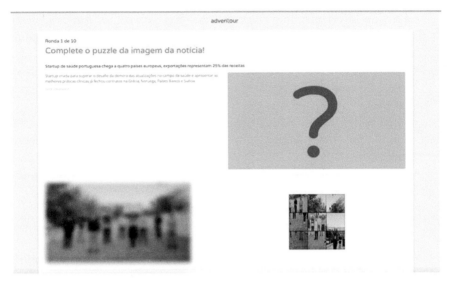

Figure 4.9 Adventour mini-game.

each journey, news and minigames (Figure 4.9) are integrated into an experience that promises to be a fun and fundamentally didactic, and encouraging curiosity for the various countries.

The game Adventour also enables players to share traveling tips, curiosities, and recommendations. A game community is also generated to exchange knowledge and traveler's experiences.

This newsgame was developed by a group of five undergraduate students in New Communication Technologies mentored by us. After designing and developing the game, user testing with five participants have also revealed the importance of the following information and functionalities: (a) information about vaccination, medication, and epidemics; (b) food and water precautions; and (c) citizen's rights relative to health and information about health reimbursements.

After playing the game, the players claimed that the game increases interest in news and everyday life matters. Also, it encourages us to take a closer look at the details of news content.

These experiences in the field and literature review on intergenerational design within a tourism setting (Vale Costa & Veloso, 2020) enables to summarize a set of recommendations to design such games:

• Run pretrip campaigns relative to the destination and e-guidance for the routes with certified information.

- Provide travel planning information—that is weather forecast, destination marketing organization's contact information, events, opening times, and accessibility.
- Recommend places and show relevant content to different points of interest based on different criteria (e.g., tourist's location, context, and profile).
- Engage the players with the place through the use quests, QR codes, follow the story initiatives, travel blogs, puzzles, and riddles, among others.
- Enable automatic check-ins, tour stamps, city keys, and sightseeing, and history clues.
- Create awareness to local resources and sustainability.
- Allow both gamer's participation (collaboration and competition) and spectatorship.
- Create a sense of "being there" with multiple narratives and navigation-challenge-rewards.

In sum, games may have an important role in extending the length of senior tourists' visits to a place and assist the player-tourist journey. The application of digital media, especially games, in health tourism has also been understudied. The next subchapter will discuss the concept of well-being and criticism toward a wellness-oriented market.

4.4 The Wellness-Oriented Market: Benefits and Shortcomings

During the past few years, there has a boom in the well-being industry—from self-care recommendations and illness prevention to personal trainers, medicine supplements, and luxury oils, and spas. However, the concern for ageing and well-being has been a humankind concern for much longer.

In philosophy, Aristotle's interest in the notion of happiness paved the way into later advancements in hedonism theories and lively discussions on people's well-being and longevity. For example, in the 18th and 19th centuries, such publications as *"An essay of health and long life"* (Cheyne, 1724) or *"Domestic Medicine"* (Buchan, 1848) introduce a culture of illness prevention and self-care.

As Cheyne (1724, p. 1) states in the Introductory part of "An Essay of Health and Long Life":

"It is a common Saying, That every Man past Forty, is either a Fool or a Physician. It might have been as justly added, that he was a Divine [i.e., clergyman] too: For, as the World goes at present, there is not any Thing that Generality of the better Sort of Mankind so lavishly and so unconcernedly throw away, as Health, except externally Felicity. Most men know when they are ill, but very few when they are well. And yet it is most certain that "it is easier to preserve Health, than to recover it; and to prevent Diseases, than to cure them."

As highlighted in this quotation, new concerns have emerged with welfare and retirement, being well-being part of a "take-care-of-yourself movement" in the industrial revolution. There is a general focus on health and longevity in this definition, however, when it comes to the concept of well-being uncertainty prevails (i.e., *Most men know when they are ill, but very few when they are well.*"). A degree of uncertainty towards the concept remains in today's society.

A multitude of definitions instead of a single one is used to characterize well-being. There are some desire and well-being theories that are enlisted in Table 4.1, but there is some criticism relative to each theory (Cooper & Kirkwood, 2014).

Some terms are related to the well-being concept (e.g., autonomy, environmental mastery, personal growth, among others; (Diener, Suh, Lucas, &

Table 4.1   Example of questioning process

Theories	Hedonism	Desire Theories	Objective List Theory
Brief description	The individual's well-being is dependent on the balance of pleasure over pain.	The individual's well-being on the satisfaction of their desires and preferences	List items that characterize well-being.
Criticism	The experience machine scenario: A repeated pleasant activity may not be ideal and not bring meaningful experiences.	Pleasurable experiences that are not part of own desire fulfilment can contribute to the individual's well-being.	Elitism (items may not be applied to an individual).

Smith, 1999; Kirkwood, Bond, May, McKeith, & Tech, 2010; Ryff & Keyes, 1995). In all, well-being encompasses both a subjective dimension (e.g., self-reported feelings) and an objective one (e.g., experiences and education throughout life) and only the merge of the two enables to understand thyself.

According to the World Health Organization (2002, p. 51), physical, mental, and social well-being are inherent in the concept of health and, hence, the wellness market is very likely to be important to foster a culture of self-care and illness prevention in society. However, caution must be also applied to this highly profitable market with false evidence advertisements and inappropriate treatments.

4.5 Concluding Remarks

This chapter introduced the concept of gamification and its interrelatedness with active learning and application in wellness-oriented markets. In this sense, some of the game elements and necessary process to game thinking and nongame contexts were clarified.

Tourism has been a negatively affected sector with the COVID-19 outbreak and as such, much attention has been directed to "slow" and rural domestic tourism. Moreover, there has been some urgency in re-stimulating in tourism offers and wellness-oriented markets.

Location-based games, Gamified apps, and Newsgames are some of the examples provided that can assist and guide the tourists' traveling pre-experience, in loco, and postexperience. Game-based approaches are very likely to help to extend the length of senior tourists' visits to a place and assist the player-tourist journey.

Finally, there has a boom in the well-being industry. If, on the one hand, this promotes preventive health and a culture of self-care, on the other hand, caution must be applied with false evidence advertisements and inappropriate treatment or medication administration.

References

Alén, E., Domínguez, T., & Losada, N. (2012). New opportunities for the tourism market: Senior tourism and accessible tourism. In M. Kasimoglu (Eds.) *Visions for global tourism industry: Creating and sustaining competitive strategies,* Rijeka, Croatia: InTech, 139–166.

Buchan (1848). Domestic medicine' A treatise on the prevention and cure of diseases by regimen and simple medicines: with observations on sea-bathing, and the use of the mineral waters. *Boston: Otis. Retrieved from* http://hdl.handle.net/2027/chi.24290817?urlappend=%3Bseq=11 (Access date: Sep 10th, 2020)

Bulencea, P., Egger R (2015). *Gamification in tourism: Designing memorable experiences.* GmbH: Books on Demand.

Butler-Kisber, L., & Poldma, T. (2010). The power of visual approaches in qualitative inquiry: The use of collage making and concept mapping in experiential research. *Journal of Research Practice, 6*(2), 1–6.

Carrol, J. M. (1999, January). Five reasons for scenario-based design. *In Proceedings of the 32nd Annual Hawaii International Conference on Systems Sciences.* 1999. HICSS-32. Abstracts and CD-ROM of Full Papers (pp. 11-pp). IEEE. DOI: 10.1109/HICSS.1999.772890

Chalanyova, O. G., & Mikulas, P. (2017). Measuring the celebrity: Contemporary Metrics of Fame. *Proceedings of the 20th International Conference "Economic and Social Development",* Prague, 27–28 April, 2017, 363–371.

Cheyne, G. (1724). An essay of health and long life. *Europeanlibraries, New York: Arno Press.* Retrieved from https://archive.org/details/anes sayhealthan00cheygoog (Access date: Sep 10th, 2020)

Cleaver, M., Muller, T. E., Ruys, H. F., & Wei, S. (1999). Tourism product development for the senior market, based on travel-motive research. *Tourism Recreation Research, 24*(1), 5–11. https://doi.org/10.1080/0250 8281.1999.11014852

Cooper, C. L., Kirkwood, T.B. L. (2014). Introduction: Wellbeing in Later Life, In R. Cooper, E. Burton, C. Cooper (Eds.) *Wellbeing: A Complete Reference Guide, Wellbeing and the Environment,* West Sussex, UK: John Wiley & Sons.

Deterding, S., Dixon, D., Khaled, R., & Nacke, L. (2011). From game design elements to gamefulness. *Proceedings of the 15th International Academic MindTrek Conference on Envisioning Future Media Environments– MindTrek* 11, 9–15. doi:10.1145/2181037.2181040

Diener, E., Suh, E. M., Lucas, R. E., & Smith, H. L. (1999). Subjective wellbeing: Three decades of progress. *Psychological bulletin, 125*(2), 276–302. DOI: 10.1037/0033-2909.125.2.276

Elias P, Rajan N O, McArthur K, Dacso C C (2013). InSpire to promote lung assessment in youth: Evolving the self-management paradigms of young people with asthma. *Medicine 2.0, 2*(1). http://doi.org/10.2196/ med20.2014

Eurostat—Statistics Explained (2018). *Tourism trends and ageing.* Retrieved from https://ec.europa.eu/eurostat/statistics-explained/index.php/Tourism_trends_and_ageing#Seasonal_patterns (Access date: September 4, 2020)

Foxman, M. H. (2015). *Play the news: Fun and games in digital journalism.* New York: Columbia University Graduate School of Journalism.

Fullagar, S., Markwell, K., Wilson, E. (2012). Starting Slow: Thinking Slow Mobilities and Experiences. In S. Fullagar, K. Markwell, & E. Wilson (Eds.) *Slow tourism: Experiences and mobilities* (Vol. 54). Bristol, UK: Channel View Publications.

Hunicke, R., LeBlanc, M., & Zubek, R. (2004, July). MDA: A formal approach to game design and game research. *In Proceedings of the AAAI Workshop on Challenges in Game AI* (Vol. 4, No. 1, p. 1722). Retrieved from https://www.aaai.org/Papers/Workshops/2004/WS-04-04/WS04-04-001.pdf (Access date: September 4, 2020)

Hunter–Jones, P., & Blackburn, A. (2007). Understanding the relationship between holiday taking and self-assessed health: an exploratory study of senior tourism. *International Journal of Consumer Studies, 31*(5), 509–516. https://doi.org/10.1111/j.1470-6431.2007.00607.x

Kapp, K. M. (2012). *The gamification of learning and instruction: game-based methods and strategies for training and education.* New Jersey, USA:John Wiley & Sons.

Kaplan, M. (1986). Leisure and aging: an international perspective. *World Leisure & Recreation, 28*(2), 6–10.

Kapp K M (2012). *The gamification of learning and instruction: game-based methods and strategies for training and education.* New Jersey, USA: John Wiley & Sons.

Karev, I., & Doron, I. (2017). The human right to leisure in old age: Reinforcement of the rights of an aging population. *Journal of aging & social policy, 29*(3), 276–295. https://doi.org/10.1080/08959420.2016.1261388

Kirkwood, T., Bond, J., May, C., McKeith, I., & Teh, M. M. (2010). Mental capital and wellbeing through life: Future challenges. In C. L. Cooper, J. Field, U. Goswami, R. Jenkins, & B. J. Sahakian (Eds.) *Mental capital and wellbeing.* New Jersey, USA: Wiley-Blackwell, 3–53.

Lieux, E. M., Weaver, P. A., & McCleary, K. W. (1994). Lodging preferences of the senior tourism market. *Annals of Tourism Research, 21*(4), 712–728. https://doi.org/10.1016/0160-7383(94)90079-5

Malone, T. W. (1980, September). What makes things fun to learn? Heuristics for designing instructional computer games. In Proceedings of the 3rd

ACM SIGSMALL symposium and the first SIGPC symposium on Small systems (pp. 162–169). https://doi.org/10.1145/800088.802839

McGonigal, J. (2011). *Reality is broken: Why games make us better and how they can change the world.* London, UK: Vintage.

Muller, M. (1993). PICTIVE: Democratizing the Dynamics of the Design Session. In D. Schuler, A. Namioka (Eds.) *Participatory Design— Principles and Practices.* Hillsdale, New Jersey: Lawrence Erlbaum Associates, Publishers.

Muller, M. J.; Druin, A. (2012). Participatory design: the third space in human–computer interaction. In Julie A. Jacko (Eds.) *Human Computer Interaction Handbook; Fundamentals, Evolving Technologies Applications. Human Factors and Ergonomics* (pp. 1125–1153), Ch.49, 3rd ed. Boca Raton: CRC Press.

Ortet, C. (2019). Gamification and Senior Cyclo-tourism: Design Proposal for the miOne Community. *Dissertação de Mestrado em Comunicação Multimédia, Universidade de Aveiro, Aveiro, Portugal.* http://hdl.handle.net/10773/27471

Ortet C.P., Costa L.V., Veloso A.I. (2019) Jizo: A Gamified Digital App for Senior Cyclo-Tourism in the miOne Community. In: Zagalo N., Veloso A., Costa L., Mealha Ó. (eds) Videogame Sciences and Arts. VJ 2019. *Communications in Computer and Information Science,* vol 1164. Springer, Cham. https://doi.org/10.1007/978-3-030-37983-4_15

Patterson, I. R. (2006). *Growing older: Tourism and leisure behaviour of older adults.* Oxfordshire, UK: Cabi.

Ryff, C. D., & Keyes, C. L. M. (1995). The structure of psychological well-being revisited. *Journal of Personality and Social Psychology,* 69(4), 719. doi: http://dx.doi.org/10.1037/0022-3514.69.4.719

Sailer, M., Hense, J. U., Mayr, S. K., & Mandl, H. (2017). How gamification motivates: An experimental study of the effects of specific game design elements on psychological need satisfaction. *Computers in Human Behavior,* 69, 371–380. doi: http://dx.doi.org/10.1016/j.chb.2016.12.033

Schell, J. (2010). *When games invade real life.* Retrieved from https://www.ted.com/talks/jesse_schell_when_games_invade_real_life (Access date: August 28, 2020)

Sellick, M. C. (2004). Discovery, connection, nostalgia: Key travel motives within the senior market. *Journal of Travel & Tourism Marketing,* 17(1), 55–71. https://doi.org/10.1300/J073v17n01_04

Shoemaker, S. (1989). Segmentation of the senior pleasure travel market. *Journal of travel research,* 27(3), 14–21. https://doi.org/10.1177/0047 28758902700304

Tiago, M. T. P. M. B., de Almeida Couto, J. P., Tiago, F. G. B., & Faria, S. M. C. D. (2016). Baby boomers turning grey: European profiles. *Tourism Management,* 54, 13–22. https://doi.org/10.1016/j.tourman.2015.10.017

United Nations General Assembly (1948). *The Universal Declaration of Human Rights.* Retrieved from https://www.un.org/en/universal-declaration-human-rights/index.html (Access date: September 9, 2020)

Vale Costa L., Veloso A.I. (2020) Gameful Tale-Telling and Place-Making from Tourists' Generation to Generation: A Review. In: Gao Q., Zhou J. (eds) *Human Aspects of IT for the Aged Population. Healthy and Active Aging. HCII 2020. Lecture Notes in Computer Science,* vol 12208. Springer, Cham. https://doi.org/10.1007/978-3-030-50249-2_47

Veloso A.I., Carvalho D., Sampaio J., Ribeiro S., Vale Costa L. (2020) Footour: Designing and Developing a Location-Based Game for Senior Tourism in the miOne Community. In: Gao Q., Zhou J. (eds) *Human Aspects of IT for the Aged Population. Healthy and Active Aging. HCII 2020. Lecture Notes in Computer Science,* vol 12208. Springer, Cham. https://doi.org/10.1007/978-3-030-50249-2_48

Werbach, K., & Hunter, D. (2012). *For the win: How game thinking can revolutionize your business.* Philadelphia, USA: Wharton Digital Press.

World Health Organisation (2002). *Active Ageing: A Policy Framework. Geneva:* World Health Organization.

You, X., & O'Leary, J. T. (1999). Destination behaviour of older UK travellers. *Tourism Recreation Research* 24, 23–24. https://doi.org/10.1080/02 508281.1999.11014854

Zichermann, G., & Cunningham, C. (2011). *Gamification by design: Implementing game mechanics in web and mobile apps.* Sebastopol, CA: O'Reilly Media, Inc.

Zichermann, G., & Linder, J. (2013). *The gamification revolution: How leaders leverage game mechanics to crush the competition.* New York, USA: McGraw Hill Education.

5

Assessing Games for Active Ageing

This chapter analyses the importance and the challenges associated with the assessment of gameplay experiences and its potential impact on individual's everyday life, especially on active ageing (i.e., health, sense of security, and participation in society; World Health Organization, 2002). After addressing the implications of game design evaluation with the users' involvement, the chapter proceeds with the presentation of different methods applied in this field and share the experience from our research group in the area and lessons learned.

5.1 Assessing the Game Experience

Assessing the gamers' motivations, interaction patterns, and the impact of the gameplay experience on everydayness and perceived sense of well-being and quality of life has been attracting quite a lot of interest lately, especially with recent advancements on positive computing (Calvo & Peters, 2014) and inter-relatedness of the use of information and communication technologies and both daily life activities and societal challenges—for example, gamification, games with a purpose, smart technology and active and assisted living.

Although the focus of gameplay evaluation has been on its impact on illness prevention and rehabilitation (Leinonen, Koivisto, Sirkka, & Kiili, 2012; Pannese, Wortley, & Ascolese, 2016), there have been quite a few studies (e.g., Allaire, McLaughlin, Trujillo, Whitlock, LaPorte, & Gandy, 2013; Baranowski, Buday, Thomson, & Baranowski, 2008; DeSmet et al., 2014; Hall et al., 2012) that have explored the effects of games on subjective health-related well-being, in comparison with nongame approaches.

In-game design evaluation, both player-centric and game-centric approaches must go altogether to ensure rewarding gameplay, sense of fairness, dilemmas, impactful storytelling, accessibility, and positive experiences that are aware of the player's context.

The Game Heuristic Evaluation method, that is the application of guidelines to assess game design has been widely used (Federoff, 2002; Pinelle, Wong, & Stach, 2008), being recognized as one of the most flexible and cheap evaluation procedures with impact on design decisions (Dix, Finlay, Abowd, & Beale, 2004).

When applying a set of heuristics to a gamer-friendly environment in an ageing society, the following categories should be taken into account:

(a) *Assistive components* refer the use of physical devices that intend to be user-customizable (adaptive) and maintain (assist), increase or improve (rehabilitate) the gamer's cognitive, emotional, social and physical capabilities and establish an interconnection with daily life.

(b) *Game elements* that embody mechanisms of help, story/game content, time commitment, and resource management, game experience and aesthetics, the outcome—reward/punishment conditions, artificial intelligence, level design, trust and security, goals/challenges, gameplay and rules, and feedback.

(c) *Usability and accessibility* that is relative to the interface and the gamer's context.

For each category, the heuristics can be summarized into the following (Veloso & Costa, 2016):

(a) *Assistive components*

(a1) *Cognitive capacity*

- Game environments should encourage self-learning.
- Consider the cognitive conditions of the player.
- Meet the learners' needs and skills.
- Change the game logic regarding the gamers' process.

(a2) *Emotional capacity*

- Stir particular emotions through sound and visual effects.
- Develop a sense of empathy between game characters and players.
- Transport the player into a level of personal involvement both emotionally and viscerally.

(a3) *Physical capacity*

- Consider the physical condition of the player.

- Create feasible goals that take into account gamers' physical condition and context.

(a4) *Social capacity*

- Support both competition and cooperation, social interactions between players, and in-game and off-game social communities.
- Provide the option of expressing players' mood and attitude statements.

(a5) *Link to daily-life activities*

- Relate the story experience to real life and hook interest.

(b) *Game elements*

(b1) *Mechanisms of help*

- Provide tutorials that mimic gameplay.
- Consider the cognitive conditions of the player.
- Meet the learners' needs and skills.

(b2) *Story/game content*

- Create an appealing storyline that encourages immersion.
- Use common and familiar themes that can be understood easily.
- Provide consistency between game elements and the story.

(b3) *Time commitment and resource management*

- Provide easy and with access to the activity of game-playing.
- Enable users to skip nonplayable and frequently repeated content.

(b4) *Game experience and aesthetics*

- Encourage fun and replay.
- Provide a positive game-based and learning experience.

(b5) *Outcome—reward/punishment*

- Reward the player by increasing their capabilities (power-ups) and expanding their skills and resources.

(b6) *Artificial intelligence (AI)*

- Provide different levels of AI depending on the level of design and players' play (novice & experts).

(b7) *Level design*
- Adjust the game challenges to the players' skills.

- Provide different levels of difficulty.

(b8) *Gameplay and rules*

- Encourage a sense of control from the player relative to their avatar, and the impact of their actions onto the game world.
- Match players' movements to on-screen actions.
- Monitor information related to the fame activity.

(b9) *Feedback*

- Provide immediate in-game feedback.
- Provide information on the player's game status and players' performance in the game.

(c) *Usability and accessibility*

- Provide consistent design, clear icons, and big fonts.
- Design an intuitive interface (avoid cognitive overload).
- Provide audio and visual (e.g., drawings, images...) representations that are easy to interpret.
- Enable players to customize video and audio settings, level of difficulty, game speed, network, environment settings, and other interface aspects.
- Minimize the number of control options and provide easy-to-use, customizable and physically comfortable controls;
- Take into account some game conventions and patterns to shorten the learning curve.
- Prevent errors and ask to confirm some decisive actions (e.g., exit the game).

Although game heuristics may be a useful tool to assess the effectiveness of the product based on previous experience in the field, the involvement of the players in the process is also important.

The past few years have seen increasing interest in the use of mixed methods to understand both the game-playing experiences (e.g., Cheng & Annetta, 2012; Fang, Lin, & Chuang, 2009; Law & Sun, 2012; Petersen, Thomsen, Mirza-Babaei, & Drachen, 2017) and the use of technologies for the aged population (e.g., Irizarry et al., 2017; Kachouie, Sedighadeli, Kosla, & Chu, 2014). This popularity is owing to the fact that the sole use of quantitative or qualitative approaches is often insufficient to understand the complexity of different aspects and lenses of a research problem (Cohen, Manion, & Morrison, 2013) that may impact both on-screen and off-screen behaviors.

Mixed-methods date back to the 1950s with a multitrait–multimethod matrix developed by Campbell & Fiske (1959) to analyze psychological traits but only very recently these have become largely used to address the research problem.

The rationale for using a mixed-method as a research design may vary and the sequence in which quantitative and qualitative approaches are applied depend on the purpose established for the study (Creswell & Clark, 2011). An example of the objectives for delineating a mixed-method design in games for active ageing, well-being, and quality of life may be:

- Develop a game addressed to older adults involving them in the process, collecting and analyzing qualitative data to assess its effectiveness to affect active ageing, well-being, and quality of life.
- Refine the game to serve as an instrument to an experimental study involving a control group or group of comparison to assess their effectiveness for active ageing.
- Cross the data obtained from both quantitative and qualitative approaches. These data will enable to assess the players' context, game strengths and weaknesses, self-perceived well-being, and quality of life, and the design/development considerations to the game industry and market.

The following sections include some of the lessons learned in carrying out interviewing techniques, experimental studies and participant observation, and in-game analysis with adult learners of the Third Age.

5.2 Surveying Techniques

A survey design can be used to assess players' perceived experience, feelings, and attitudes toward gameplay. Surveys can be administered in the form of questionnaires or interviews (Creswell & Creswell, 2018), varying in terms of purpose, sampling, and undertaken procedures. This section provides a brief overview of the differences and procedures to take in questionnaires, interviews, and group workshops.

5.2.1 Questionnaires

Questionnaires are an instrument for data collection, which purpose is often to generalize to the whole population data obtained from a random sample. In the context of games, these are used to characterize the sociodemographic profile of players and be the basis to create personas (Grudin & Pruitt, 2002).[1]

When administering a questionnaire, the population of the study needs to be defined (e.g., all Portuguese students at the University of the Third Age). Depending on the total number of the population, the next step is to determine the sample size and confidence level, that is uncertainty level willing to accept or refute a set of hypotheses.

For example:

In the SEDUCE 2.0 research project (Use of Communication and Information in the miOne online community), the population is 46061 adult learners/328 Portuguese Universities of Third Age and the margin of error to accept is 5%. The recommended sample size is, therefore, 381 adult learners/178 Universities of Third Age.[2]

Given the impossibility of compiling a list of all population members (in this case, all adult learners at the University of the Third Age), clustering is often needed to identify groups or organizations and list all those cluster members. It is worth noting, however, that the researcher needs to make it very clear the goal of the study, ethics principles, and data collection and protection procedures.

After identifying and listing the population, the probability to select one of the individuals to a certain sample must be equal and, thus, they need to be randomly selected.[3] Such sociodemographic data as age, gender, education, and location must also reflect the proportion of the population, in case of a stratified sample. Nevertheless, a convenient sample may be used in the impossibility to randomly select a sample. In these latter cases, researchers may be very cautious about not making attempts to extrapolate data and acknowledging the limitations of data to a specific group.

Depending on the specific research goals determined for assessing games, an instrument to collect the information can be designed for the research or adapted or, alternatively, use an intact version of that instrument. In these

---

[1] Personas are fictional characters that represent the user narratives (their motives, thoughts, routines, statements, feelings, and scenarios.)

[2] For estimating this value, a sample size calculation tool was used, http://www.raosoft.co m/samplesize.html (Access date: Aug. 29, 2020).

[3] Assign a specific number to each individual/organization and then generate random numbers. Start recruiting the individuals/organizations that have that generated number and in case of refusal, count them in the number of contacted but refused, and continue the whole process

two latter cases, asking permission to use it is needed from both authors and publishers.

In a letter for asking permission to adapt/use instruments for data collection, the following data is often provided:

- Provide study details and the reason the scale/survey would be relevant to accomplish the goals of the game assessment.
- Ask for license entire or partial questionnaire (in case of adaptation) to be administered and the procedures to receive reliable and valid data/results.
- Check whether there is a language version of the questionnaire validated to the country which it is intended to be administered, administration mode (web, self-administration via paper/pencil and telephone, use of interview script) and the procedures to undertake in data collection for a survey.
- Ask for any recommendation in score computing and analysis.
- Ensure that all credits will be provided in the APA form as well as an acknowledgment to the author.

There are a number of instruments that can be used to assess the player's perception toward the game activity—for example, the Consumer Video Game Engagement (Abbasi et al., 2019), Perceived-Challenge Gameplay Interaction (Denisova, Cairns, Guckelsberg, & Zendelle, 2020), Play Experience (Pavlas et al., 2012), among others. For choosing the instrument to administer, one should take into account the following elements: purpose, internal consistency, language validation and localization, and permission given by the authors.

In regard to survey design for the research, the information type to be gathered needs to be defined (e.g., biographical information, behaviors, routines, attitudes). According to Punch, (2005), questionnaires may be divided into the following pieces of information: (a) cognitive/knowledge; (b) affect; and (c) behavior.

The same author (Punch, 2005) highlights that delineating the types of variables, dimensions, and measures is key to develop a good questionnaire. The variables tend to be associated with hypotheses testing supported in the literature review.

For example:

In our previous research (Costa & Veloso, 2016, p. 38) that aims to examine the motivations and routines of gamers aged 50 and over (motivations, what type of games and video games they play, frequency of playing, and how comfortable they feel with different technologies), the following hypotheses were elaborated:

H_1: There is a correlation between the types of games participants like to play and their age;

H_2: There is a correlation between who participants live with and the modes of game-playing;

H_3: There is a correlation between the type of games preferred and the skills that gamers want to practice.

After defining the variables and hypotheses, the questions (survey items) are elaborated and can be pilot tested. Pilot testing will enable us to assess the following criteria:

- The scope and the purpose of the research is (not) explained.
- The instructions are (not) sufficiently and clearly described.
- There was (not) risk of multiple interpretations or ambiguity in the use of terms.
- The sequence of the questions.
- Suitability of the questions to each section.
- Number of repeated or inappropriate questions.
- Terms that need to be explained.
- Adequacy of the scale used.
- Suitability of the questions formulated for the analysis.
- Validity of the content in scientific terms.

Finally, a timeline for administering the survey and analysis is followed. The researcher needs to document the number of returns of the questionnaire, the method used to determine response bias and depending on the sampling technique used and robustness of the survey, provide a descriptive analysis (e.g., number and percentage of the participants, average and standard deviation, and correlation between variables) or inferential (e.g., cross-tabulation and chi-square tests; independent *t*-test).

5.2.2 Interviews

Surveying can also occur in the form of interviews, either "face-to-face," telephone, video chat, or text-based messages. They may rely on the use of predetermined questions listed in an interview script with little variation (structured) or with more flexibility (and semi-structured) or overly dependence on the interviewee's answers (unstructured; Gill et al., 2008).

During the recruitment process, the number of invitations sent out, acceptances, refusals, and nonanswers should be also documented. Relative to the recommended procedures to be undertaken when interviewing participants, Rose (1994) highlights the following:

(a) Choose a set with minimal or no distractions.
(b) Inform the interviewee about the purpose of the interview, the reason(s) for being selected to the interview, duration, and confidentiality terms.
(c) Put the interviewee at ease.
(d) Be attentive to the procedures for data collection and field notes.

Based on the interviewing procedures, interviewee's information, and the research goal, an interview protocol is created.

The protocol is usually divided into the following steps:

1. Introduction/instructions and standard procedures
2. Ice-breaker questions
3. 4/5 questions
4. Thank-you statements

In the introduction/instructions and standards procedures (1), researchers present themselves, the aim of the research, and the reason for being invited for the research. In this step, the interviewee is also informed about the duration of the interview, the procedures undertaken for data collection, and data anonymity. Asking permission to audio-record the interview for subsequent transcription may be also included and the interviewee should be recalled that there are no correct or wrong answers for the questions, just their point of view.

Then, one or two ice-breaker questions (2) should be considered. These would be useful to help to characterize the interviewees—for example,

Maybe you could introduce yourself and say a few words about your experience as a gamer; How long did you begin to play game X, Y?

Interview questions should not exceed five, considering the interviewee's time availability and risk of fatigue (maximum duration of 60 minutes), and each question posed must have an inherent analysis goal.

Finally, the interview should end with a thank-you statement. The interviewer could ask whether there is anything the interviewee would like to add or talk about that has not been covered, thank the participation and be at the disposal for additional contact.

5.2.2.1 Contextual inquiry

When assessing psychology and sociability issues associated with the game-play activity, a contextual inquiry is likely to be used. The purpose of the contextual inquiry is to assess the users' practices and experiences with an artifact in a natural setting (Holtzblatt, Wendell, & Wood, 2004; Namioka & Schuler, 1993)—for example, homes, Universities of Third Age, retirement homes, and day care centers.

One-to-one interviews tend to be performed in contextual inquiry with observations. Three principles guide the process: (a) be context-oriented; (b) reinforce the dialogue between the designer and the user (partnership and interpretation); and (c) attend the focus of the study (Holtzblatt et al., 2004; Namioka & Schuler, 1993).

Surveying is one of the methods that can be used in contextual inquiry to assess the users' preferences and motivations, alongside operability assessment whether a simulated interface is being tested; brainstorming product ideas; collecting context-based information through field visits and observations (e.g., photographs or cultural probes) and/or group discussions and interviews (Namioka & Schuler, 1993). The following section provides a brief overview of the use of group interviews to gather context-based data.

5.2.2.2 Group interviews

Group interviews are a variant of these interviews that beyond gathering information about the player, they enable to analyze the interactions among the participants.

These interviews often involve a small group of 5–7 participants and a skilled moderator (Debus, 1995, p. 8). According to Morgan (1997), group interviewing assumes the following functions: (a) be the main source of data collection; (b) support data from other sources (e.g., survey); and (c) be part of a mixed-method approach.

Considering the popularity of computer-mediated communication (i.e., video-based and *instant messaging*), group interviews have also incorporated these online tools into the research practices (Rezabek, 2000). Indeed, online group discussions have the potential to involve players with different cultures, contexts, and backgrounds, enriching the knowledge about different user's experiences.

Table 5.1 shows the advantages and disadvantages of online group discussions, in comparison with those that are performed face-to-face (Rezabek, 2000).

Table 5.1 Comparison between face-to-face and online group discussions

	Face-to-Face Discussions	Online Group Discussions
Advantages	–Visual cues/facial expressions and nonverbal communication can be easily assessed.	–Reach a broad physically-distance geographic audience.
		–Easier to schedule the discussion, speeding up the process.
	–Traditionally used in Academic Research.	–Participants may be comfortable in their natural settings (e.g., home, workplaces) and may ensure anonymity.
Disadvantages	–Limit the heterogeneity of the participants and there is a risk of member bias (homogeneity in culture, place, and context)	–Difficult to manage different time zones.
		–In case of using discussion boards, they may be extended over time, making data analysis an arduous task.
		–Digital illiterate users are likely to be excluded.
When to use	–When nonverbal communication and observation play a significant role.	–Involve participants from different geographic places.

As illustrated in Table 5.1, online group discussions are often employed when involving participants from different geographic places. For that, both synchronous and asynchronous communication services may be used. Whereas participants may participate synchronously (e.g., chat rooms), asynchronous group communication can be done through the use of email, list servers, mailing lists, and discussion groups.

In the context of games, group interviews are a good mechanism to meet the following goals: (a) generate new ideas for a game; (b) validate the relevancy of game concepts; and (c) provide insights into the design of a game proof concept. In fact, they should be used as an innovation source for generating game ideas and concepts, instead of game validation (Fullerton, Swain, & Hoffman, 2008).

As for the involvement of older adults in these group interviews, such recommendations as limiting the topics to cover, and duration (Pak & McLaughlin, 2010) should be added to the recommended procedures to undertake group interviews. In brief, these are the following:

- Limit the number of topics to cover (between 5 and 7).
- Limit the hour session to a maximum of 90 minutes.
- Create an environment which promotes the participants' shareability of their gameplay experiences. Some participants do not feel comfortable to admit that they play games because they had a bad connotation over time and there is some prejudice against older people who play games.
- Manage intra-group conflicts (Gallagher, 2005) that may arise and reassure that all data will be kept confidential and that there are no right and wrong answer. The moderator should also be sensitive to some questions that may bring up emotions related with a sense of well-being and quality of life and respect the participant's right to not answer or withdrawal anytime.
- Recruit a heterogeneous group to these interviews—for example, gender-balanced, different levels of gameplaying experience (novice, hard-core, moderate, and casual);
- Make sure to introduce game concepts and conventions at the beginning of the interview, especially when involving novice players.
- In online group interviewing, threading and "netiquette" issues are some aspects to additionally consider (Oringderff, 2008);
- Structure these group interviews into the following: Welcome message, one or two ice-break questions, Overview of the topic/introduction of game concepts, purpose, rules, questions, and exercises (Doyle, 2009).

Data analysis proceeds to data collection of the interviews. In data analysis, verbatim transcriptions and/or researcher notes that may include videos, audio, photographs, and text are compiled, re-read, and coded into categories.

Two coding processes are usually applied:

- *Pre-determined coding:* These codes are established a *priori* the coding process based on the literature review or previous expectations relative to the interrelatedness of the variables.
- *Open coding:* These codes emerge from the data obtained from the participants. This coding process relies on Grounded Theory (Glaser & Strauss, 2006) that supports the idea that given the challenges of changeable phenomena (see, for example, game trends), codes are generated from the data obtained *in situ* and at the moment.

Hand-coding can be a very time-consuming process but there are a number of computer-assisting data analysis software (e.g., QSR NVivo, WebQDA) that enables the user to easily locate different pieces of information (i.e., verbatim transcriptions, memos, photographs, and videos), assign codes, and query the

Table 5.2 Example of questioning process

Questions	Qualitative Data Tools	Search Restriction
What are the interviewees' favorite games?	Search for frequent words	Only the category "Favorite games"
Is there any relationship between the type of difficulties felt while playing games and the participants' age?	Data matrix	Categories Age group Type of difficulties (accessibility, interaction, language used)
Do interviewees who are against playing in later life base their opinions on previous gameplay experiences or reasons for not playing?	Compound query	Categories: being against and previous gameplay experience; being against and reasons

relationship between codes. For data analysis, the procedures presented below are recommended:

1. (Re) read all the documents and assign an equal ID number to documents and the source of information.
2. Invite the interviewees to review the transcriptions.
3. Highlight the participants' statements and compare them, interrelating information.
4. Code the information (predetermined coding or open coding).
5. Provide a codebook with the code, code description, number of references coded, and exemplar quotes that were assigned to that code.
6. Associate data collection questions to data analysis questions and query data, using such tools as search frequent words, compound queries, or effects matrix. Table 5.2 illustrates some of the tools used in the data questioning process.
7. Ensure data validity and reliability in data analysis through, for example, the use of member checking or triangulation of data.

This subchapter provided a brief overview of surveying techniques. The next subchapter will explain the relevancy and considerations in using experimental studies to assess gameplay experiences and A/B testing.

5.3 Experimental Studies

User testing relies on many experimental studies, especially when assessing the effectiveness of technology-mediated interventions. In specific, a group-comparison experimental study often involves different types of stimuli, random assignment of participants to different groups, and observable changes before and after an experiment or nonexperiment. Similar to questionnaire surveying, participants may be randomly selected to a sample but what characterizes a design study as a true experiment is the random assignment of participants to a group.

As aforementioned in the Subchapter 5.2., random selection refers to the equal probability of a member of the population to be assigned to a different group. Random assignment is relative to an equal probability of the participants being assigned to each group (i.e., experimental group, group of comparison, and control group) or to a certain order of the experience administration (i.e., counterbalancing).

When assessing users' interactions with a certain product or interface, experimental studies are likely to be used in summative evaluations, that is comparisons of products or interfaces relative to the competition (Sauro & Lewis, 2012).[4]

Considering that a cause-and-effect relationship between dependent and independent variables (see Subchapter 5.2) is associated with this study design, the study purpose, and a number of hypotheses are formulated before advancing to group comparison.

The procedures to carry out an experimental study can be summarized as follows:

1. Define the purpose of the evaluation and game artifact/interface comparison, *for example, Compare the effectiveness of place collectibles in increasing time average walking in comparison with skill-based badges in a gamified system; Compare the effectiveness of game-based learning to develop a positive perception of active ageing, in comparison with an online course.*

2. Formulate the hypotheses associated with a possible cause-and-effect between independent and dependent variables. For example:

   H_1: The mean of walking time participants using place collectibles is greater than the mean of participants using skill-based badges.

   H_0: There is no difference between the mean walking time of participants using place collectibles and the mean of participants using skill-based badges.

---

[4]Further information on summative valuation and the difference between formative and summative evaluations can be found on the NN/g Nielsen Norman Group website, https://www.nngroup.com/articles/formative-vs-summative-evaluations/ (Access date: Aug. 15, 2020).

3. Depending on the purpose of the study and the formulated hypotheses, the type of experiment (e.g., pretest-posttest control-group, post-test-only control-group, solomon focus-group, A-B-A single-subject) to be undertaken is identified.
4. Randomly select a sample from the population and randomly assign the participants from the sample to each experimental, comparison, or/and a control group.
5. Administer the measures of the dependent variable—for example, mean of walking time, frequency increase, attention span, attitude toward the ageing process.
6. Administer the game-based or interface intervention to the experimental group (receives the intervention/treatment) and an alternative (may receive a different intervention/treatment) or no experience to a control group (without the intervention). The use of a control group may also help to control extraneous variables (outer aspects that may influence the results obtained).
7. Extend the administration of the intervention/experience for a considerable length of time to observe eventual changes in the values of the dependent variables and control extraneous variables.
8. Compare the performance of each group relative to the dependent variables, using tests of statistical significance.
9. Analyze and interpret the data, acknowledging possible threats to the internal and external validity of the study. Internal validity is relative to the unambiguity that there is in the casual-effect relationship (or lack of it) established between the independent and dependent variables (Girden & Kabacoff, 2011). External validity refers to external aspects that may compromise the generalization of the results.

Some of the threats to internal validity are history (external factors occurring during the experiment may affect the results); maturation (participants may change/mature during the experience); regression (participants who score extremely high or low in pre-test); implementation and instrumentation (give the experimental group special treatment that may affect the results); selection (risk of involving participants with certain characteristic or personality trait, creating bias); location (the place in which the experimental study is conducted may influence the results); mortality (participants' death or dropout during the experience) and testing (participants may be familiar with the pretest outcomes and affect the responses in post-test).

In the case of testing game-playing for active ageing and healthy lifestyles, special caution is needed with the mortality given the participants' death in the course of the experiment. Another common threat is related to history, for example, life events that may influence the participants' attitude towards the ageing process or previous knowledge in a certain game.

In terms of data analysis, pretest and posttest observation measures are described (means, standard deviations, and ranges) and inferential statistical tests are used to test the hypotheses. The choice of the inferential statistical test will depend on the type of independent and dependent variables (i.e., categorical, numeric or ordinal) and the number of groups.[5] Finally, hypotheses testing is verified, that is supported and refuted, and reasons are presented based on the literature review.

The next subchapter covers the procedures for data observation and the use of eye-tracking and in-game data.

5.4 Participant Observation and In-Game Data

Participant observation and game logs are good instruments for capturing in-the-moment happenings, being data collected either from the respondents' statements or direct and indirect observation (Quivy & Van Campenhoudt, 2017).

An observation protocol is usually used for recording the design sessions it may contain the following elements: reference number, place, date and time, activity and goals, portraits/description of the main actions with the participants' statements, and references to audio-visual materials, photos, and documents.

Observers-researchers have a crucial role in collecting, interpreting, and attributing meaning to data. Their level of experience is important to conduct group discussions, collect information, and analyze data. By contrast, the relationship established between participants and the researcher can bring additional bias to the study. Hence, the researcher's role needs to be clarified.

Game logs also enable us to monitor game events and assess interface navigation, number of times, and effort to hit a game target, playing time, and perceived time.

Eye-tracking is a method that relies on the recording of eye-movement and gaze movements when presenting different stimuli (Ehmke & Wilson, 2007). The corneal reflection and pupil dilation are tracked, unraveling

---

[5]For further information on the choice of the statistical test, we recommend the Howell's book "Statistical Methods for Psychology" (Howell, 2013).

what draws the player's attention and distractions beyond self-reporting data (Hussain et al., 2018). For that, eye-trackers are used.

Eye-tracker has become very popular in the human—computer interaction field since 1947 hallmarked by Paul Fitts's study in eye-tracking cockpits interface (Jacob & Karn, 2003). The players' gaze patterns are tracked through the use of an eye-tracker, that is, it emits infrared light, recording the light reflection in the retina and track the pupil center in relation to the corneal reflections. That way, the eye-tracker enables to monitor the players' eye fixations (staring behavior) and saccades (rapid eye movements to avoid blur).

In the context of games, eye-tracking can be used to assess the players' visual behavior—that is attention, search, navigation, "point of regard," and spatial orientation in the presence/absence of certain game elements (Almeida, Mealha and Veloso 2016; Kenny et al., 2005; Sennersten, 2004).

When assessing the older adult target group, the researcher has to take into account the eye-tracker sensitivity to detect the participants' eye movements, given the effects of aging (Carter, Obler, Woodward, & Albert, 1983; Spooner, Sakala, & Baloh, 1980) and age-related macular degeneration (Pak & McLaughlin, 2010). This has been one of the greatest challenges for our master's students evaluating prototypes.

Figure 5.1 illustrates the evaluation of Jizo, a gamified app for Senior Cyclotourism (Ortet, Costa & Veloso, 2019), in which user testing in a computer had to be used as an alternative to the use of the mobile setup, owing to the fact that there was some difficulty in recognizing older adults' gaze.

Figure 5.1 (Continued)

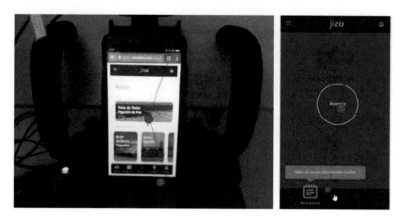

Figure 5.1   User testing of the Jizo App (Ortet et al., 2019).

Alongside this aforementioned challenge, the researchers/testers should notice whether these participants use glasses, their previous experience with digital devices, task understanding, and familiarity with game conventions.

5.5 Concluding Remarks

This chapter addressed the rationale, procedures and challenges associated with the assessment of gameplay experiences and its potential impact on an individual's everyday life, especially on active ageing.

Both player-centric and game-centric approaches need to be combined to assess a gameplay experience and this can be done through heuristic evaluation, surveying, contextual inquiry, group-comparison experimental studies, participant observation, and in-game data.

The advantages and application scenarios of each method have been discussed and so do some common threats that researchers deal with in user testing are related with mortality and history.

In sum, a mixed-method research design is key to triangulate different type of data and (re) design game-based fun and memorable experiences.

References

Abbasi, A. Z., Ting, D. H., Hlavacs, H., Costa, L. V., & Veloso, A. I. (2019). An empirical validation of consumer video game engagement: A playful-consumption experience approach. *Entertainment Computing, 29,* 43–55. https://doi.org/10.1016/j.entcom.2018.12.002

Allaire, J. C., McLaughlin, A. C., Trujillo, A., Whitlock, L. A., LaPorte, L., Gandy, M. (2013). Successful Aging Through Digital Games: Socioemotional Differences Between Older Adult Gamers and Non-gamers. *Computers in Human Behavior,* 29 (4): 1302–1306. doi:10.1016/j.chb.2013.01.014.

Almeida, S., Mealha, O., & Veloso, A. (2016). Video game scenery analysis with eye tracking. *Entertainment Computing,* 14, 1–13. https://doi.org/10.1016/j.entcom.2015.12.001

Baranowski, T., Buday, R., Thompson, D. I., Baranowski, J. (2008). Playing for Real. Video Games and Stories for Health-Related Behavior Change. *American Journal of Preventive Medicine,* 34 (1), 74–82.e10. doi:10.1016/j. amepre.2007.09.027.

Calvo, R. A., & Peters, D. (2014). *Positive Computing: Technology for Wellbeing and Human Potential.* Cambridge, Massachusetts: The MIT Press.

Campbell, D. T., & Fiske, D. W. (1959). Convergent and discriminant validation by the multitrait-multimethod matrix. *Psychological Bulletin,* 56(2), 81–105. doi: http://dx.doi.org/10.1037/h0046016

Carter, J. E., Obler, L., Woodward, S., & Albert, M. L. (1983). The effect of increasing age on the latency for saccadic eye movements. *Journal of Gerontology,* 38(3), 318–320. DOI: 10.1093/geronj/38.3.318

Cheng, M. T., & Annetta, L. (2012). Students' learning outcomes and learning experiences through playing a Serious Educational Game. *Journal of Biological Education,* 46(4), 203–213. https://doi.org/10.1080/00219266.2012.688848

Cohen, L., Manion, L., & Morrison, K. (2013). *Research methods in education.* New York, USA: Routledge.

Costa, L. V., & Veloso, A. I. (2016). Factors influencing the adoption of video games in late adulthood: a survey of older adult gamers. *International Journal of Technology and Human Interaction (IJTHI),* 12(1), 35–50. DOI: 10.4018/IJTHI.2016010103

Creswell, J., & Creswell, J. D. (2018). *Research Design: qualitative, quantitative, and mixed methods approaches (5th ed.).* Thousand Oaks, California: SAGE Publications, Inc.

Creswell, J. W., Klassen, A. C., Plano Clark, V. L., & Smith, K. C. (2011). Best practices for mixed methods research in the health sciences. Bethesda (Maryland): National Institutes of Health, 2013, 541–545.

Debus, M. (1995). Methodological review: A handbook for excellence in focus group research. Washington, D.C.: Academy for Educational Development

Denisova, A., Cairns, P., Guckelsberger, C., & Zendle, D. (2020). Measuring perceived challenge in digital games: Development & validation of the challenge originating from recent gameplay interaction scale (CORGIS). *International Journal of Human-Computer Studies,* 137, 102383. https://doi.org/10.1016/j.ijhcs.2019.102383

DeSmet, A., Van Ryckeghem, D., Compernolle, S., Baranowski, T., Thompson, D., Crombez, G., De Bourdeaudhuij. I. (2014). A Meta-analysis of Serious Digital Games for Healthy Lifestyle Promotion, *Preventive Medicine,* 69: 95–107. doi:10.1016/j.ypmed.2014.08.026.

Dix, A., Finlay, J., Abowd, G.D., Beale, R. (2004). *Human-computer interaction,* 3rd ed., Essex, England: Pearson Education Prentice Hall.

Doyle, J. (2009). Using focus groups as a research method in intellectual disability research: a practical guide. *The national federation of voluntary bodies providing services to people with intellectual disabilities.* Retrieved April, 5, 2010 from http://www.fedvol.ie/_fileupload/Research/focus%2 0groups%20a%20practical%20guide.pdf

Ehmke, C., & Wilson, S. (2007). Identifying web usability problems from eye-tracking data. People and Computers XXI HCI. But Not as We Know It—*Proceedings of HCI 2007: The 21st British HCI Group Annual Conference,* 3–7 September 2007. https://doi.org/10.14236/ewic/hci2007.12

Fang, K., Lin, Y., & Chuang, T. (2009). Why do internet users play massively multiplayer online role–playing games? *Management Decision,* 47(8), 1245–1260. doi:10.1108/00251740910984523

Federoff, M.A. (2002). *Heuristics and usability guidelines for the creation and evaluation of fun in video games [PhD thesis],* Indiana, USA: Indiana University. Retrieved from https://citeseerx.ist.psu.edu/viewdoc/downlo ad?doi=10.1.1.89.8294&rep=rep1&type=pdf (Access date: August 26, 2020)

Fullerton, T., Swain, C., & Hoffman, S. (2008). *Game design workshop: a Playcentric approach to creating innovative games,* 2nd ed., Boca Raton, FL: CRC Press.

Gallagher, P. (2005). Synchronous computer mediated group discussion. *Computers Informatics Nursing,* 23(6), 330–334.

Gill, P., Stewart, K., Treasure, E., & Chadwick, B. (2008). Methods of data collection in qualitative research: interviews and focus groups. *British Dental Journal,* vol. 204, 291–295, https://doi.org/10.1038/bdj.2008.192

Girden, E., Kabacoff, R. (2008). *Evaluating Research Articles From Start to Finish.* 3rd ed. Thousand Oaks: SAGE Publications Inc.

Glaser, B. G., & Strauss, A. L. (2006). *The Discovery of Grounded Theory: Strategies for Qualitative Research.* New Jersey, USA: Transaction Publishers.

Grudin, J., & Pruitt, J. (2002). Personas, Participatory Design and Product Development: An Infrastructure for Engagement. *Proceedings of Participation and Design Conference (PDC2002),* Sweden, 144–152.

Hall, A. K., E. Chavarria, V. Maneeratana, B. H. Chaney, and J. M. Bernhardt. (2012). Health Benefits of Digital Videogames for Older Adults: A Systematic Review of the Literature. Games for Health Journal, 1 (6): 402–410. doi:10.1089/ g4 h.2012.0046.

Holtzblatt, K., Wendell, J.B., & Wood, S. (2004). *Rapid Contextual Design: A How-to Guide to Key Techniques for User-Centered Design.* San Francisco, CA: Elsevier.

Howell, D. (2013). *Statistical Methods for Psychology.* 8th ed. Belmont, CA: Wadsworth Cengage Learning.

Hussain, J., Khan, W. A., Hur, T., Bilal, H. S. M., Bang, J., Ul Hassan, A., Afzal, M., & Lee, S. (2018). *A multimodal deep log-based user experience (UX) platform for UX evaluation.* Sensors (Switzerland), 18(5). https://doi.org/10.3390/s18051622

Irizarry, T., Shoemake, J., Nilsen, M. L., Czaja, S., Beach, S., & DeVito Dabbs, A. (2017). Patient Portals as a Tool for Health Care Engagement: A Mixed-Method Study of Older Adults With Varying Levels of Health Literacy and Prior Patient Portal Use. *Journal of Medical Internet Research,* 19(3), e99. doi:10.2196/jmir.7099

Jacob, R. J., & Karn, K. S. (2003). Eye tracking in human-computer interaction and usability research: Ready to deliver the promises. In R. Radach, H. Deubel, J. Hyona (Eds.) *The Mind's Eye: Cognitive and Applied Aspects of Eye Movement Research,* Amsterdam, The Netherlands: Elsevier Science.

Kachouie, R., Sedighadeli, S., Khosla, R., & Chu, M.-T. (2014). Socially Assistive Robots in Elderly Care: A Mixed-Method Systematic Literature Review. *International Journal of Human-Computer Interaction,* 30(5), 369–393. doi:10.1080/10447318.2013.873278

Kenny, A., Koesling, H., Delenay, D., McLoone, S. & Ward, T. (2005). A preliminary investigation into eye gaze data in a first person shooter game. *Proceedings of the 19th European Conference on Modelling and Simulation.* Retrieved August 26, 2020 from https://pdfs.semanticscholar

.org/5a0e/faf4d3a944afa99a1c0dcba2a4983e111a3d.pdf?_ga=2.25753146
4.271796822.1599047274-1345335481.1582998903

Law, E. L. C., & Sun, X. (2012). Evaluating user experience of adaptive digital educational games with Activity Theory. *International Journal of Human-Computer Studies*, 70(7), 478–497. doi: https://doi.org/10.1016/j.ijhcs.2012.01.007

Leinonen, M., Koivisto, A., Sirkka, A., Kiili, K (2012). Designing Games for Well-being; Exergames for Elderly People. *In Proceedings of the 6th European Conference on Games-based Learning, ECGBL 2012*, Cork, Ireland, October 4–5, 2012, 635–640.

Morgan, D. (1997). Focus Groups as Qualitative Research, 2nd ed., Thousand Oaks, USA: SAGE Publications, Inc.

Namioka, A., & Schuler, D. (1993). *Participatory design: Principles and Practices*. Boca Raton, FL: CRC Press.

Oringderff, J. (2008). "My Way": Piloting an Online Focus Froup. The International Journal of Qualitative Methods, 3(3), 69–75. https://doi.or g/10.1177/160940690400300305

Ortet C.P., Costa L.V., Veloso A.I. (2019) Jizo: A Gamified Digital App for Senior Cyclo-Tourism in the miOne Community. In: Zagalo N., Veloso A., Costa L., Mealha Ó. (eds) *Videogame Sciences and Arts*. VJ 2019. Communications in Computer and Information Science, vol 1164, Springer, Cham. https://doi.org/10.1007/978-3-030-37983-4_15

Pak, R., & McLaughlin, A. (2010). *Designing displays for older adults*. Boca Raton, FL: CRC Press Taylor & Francis Group

Pannese, L., Wortley, D., Ascolese. A. (2016). Gamified Wellbeing for all Ages-How Technology and Gamification can Support Physical and Mental Wellbeing in the Ageing Society, In: E. Kyriacou, S. Christofides, C. Pattichis (eds.) *XIV Mediterranean Conference on Medical and Biological Engineering and Computing 2016*, IFMBE Proceedings, Vol. 57, 1281–1285. Cham: Springer. doi:10.1007/978-3-319-32703-7_246.

Pavlas, D., Jentsch, F., Salas, E., Fiore, S. M., & Sims, V. (2012). The play experience scale: development and validation of a measure of play. *Human factors*, 54(2), 214–225. Doi: 10.1177/0018720811434513

Petersen, F. W., Thomsen, L. E., Mirza-Babaei, P., & Drachen, A. (2017). Evaluating the Onboarding Phase of Free-to Play Mobile Games Games: A Mixed-Method Approach. *CHI PLAY'17 Proceedings of the Annual Symposium on Computer-Human Interaction in Play*, 377–388. doi:10.1145/3116595.3125499

Pinelle, D., Wong, N., & Stach, T. (2008). Heuristic evaluation for games: usability principles for game design. *CHI'08 Proceedings of the SGCHI Conference on Human Factors in Computing Systems,* April 5–10, 2008, Florence, Italy, 1453–1462. https://doi.org/10.1145/1357054.1357282

Punch, K. F. (2005). *Introduction to social research: Quantitative and qualitative approaches.* 3rd ed., London: SAGE Publications Ltd.

Quivy, R., & Van Campenhoudt, L. (2017). *Manual de investigação em ciências sociais.* Lisboa: Gradiva

Rezabek, R. J. (2000). Online focus groups: Electronic discussions for Research, *Forum Qualitative Sozialforschung/Forum: Qualitative Social Research,* 1(1), Art. 18, http://nbn-resolving.de/urn:nbn:de:0114-fqs0001185.

Rose, K. (1994). Unstructured and semi-structured interviewing, *Nurse Researcher,* 1(3), 23–32. doi: 10.7748/nr.1.3.23.s4

Sauro, J., & Lewis, J.R. (2012). Comparing Completion Rates, Conversion Rates, and A/B Testing. *In Quantifying the User Experience: Practical Statistics for User Research.* Waltham, MA, USA: Elsevier.

Sennersten, C. (2004). *Eye movement in an action game tutorial.* Retrieved September 2, 2020, from http://www.simge.metu.edu.tr/journal/eyetrackin g.pdf

Spooner, J. W., Sakala, S. M., & Baloh, R. W. (1980). Effect of aging on eye tracking. *Archives of Neurology,* 37(9), 575–576. DOI: 10.1001/arch-neur.1980.00500580071012

Veloso, A. I., & Costa, L. V. (2016). Heuristics for designing digital games in assistive environments: Applying the guidelines to an ageing society. *2016 1st International Conference on Technology and Innovation in Sports, Health and Wellbeing (TISHW),* 1–3 Dec. 2016, Vila Real, Portugal, IEEE, doi:10.1109/tishw.2016.7847789

WHO (2002). Active Ageing: A Policy Framework. Second UN World Assembly on Ageing, Madrid, Spain. Retrieved from https://extranet.w ho.int/agefriendlyworld/wp-content/uploads/2014/06/WHO-Active-Agei ng-Framework.pdf (August 1st, 2021)

6

Conclusion

This book set out to discuss the way games can be designed and used for active ageing and healthy lifestyles. Considering the relevancy to the coverage of a transdisciplinary approach (e.g., knowledge in information and communication sciences, gerontechnology, marketing, among other fields), it provided deeper insights into the game market and silver economy; design of game-based tools for active ageing, gamification, senior tourism and the wellness market; and game assessment for active ageing.

"The Game Market and Silver Economy" unraveled the popularity of the "silver market" given the increasing aging market and subsequent need to leverage products and services that meets the users' needs, motivations, routines and mobility. For that, age-neutrality is important to avoid age bias and stereotypes when addressing products to the target group.

A more detailed account of a framework that constitutes the different dimensions of gereontechnology is given and *brain teasing* seems to be the major motivation of older adult players to game-play.

Chapter "Designing Game-based tools for Active Ageing" has paved the way into great advanced knowledge in design and development of these games. Although there are many marketed-oriented applications entitled "for active ageing," the other dimensions that go beyond health and define the concept of active ageing are often overlooked, namely the sense of security and participation in society.

Another discussion that emerged in this Chapter was the fact that the move of human–computer interaction toward a much more humanistic approach has brought to the fore the need to embody more and more socio-technical dynamics and community-centered design, instead of solely relying on the cognitive model of the player.

Moreover, games were highlighted to have potential to constitute a non-pharmacological intervention and examples of their use in rehabilitation and brain training were illustrated.

In chapter "Gamification, Senior Tourism and the Wellness Market," the use of game elements and techniques in Senior Tourism has been shown very promising. In times of pandemics, the tourism has been negatively affected and, as such, new strategies for encouraging "slow" and rural domestic tourism are much more in need.

In specific, gamification (use of game elements and techniques in contexts that go beyond entertainment purposes) may play a key role in supporting decision-making and adherence to a sustainable "slow tourism" attitude. In fact, game elements and challenges in unexplored places may augment the tourists' historical and cultural experiences with challenges *in situ,* and reinforce a sense of attachment with the visited place.

Previous experiences in the design and development of games addressed to Senior Tourism (i.e., a Location-based game for senior tourism, a gamified app for senior cyclotourism, and a newsgame about health tourism and epidemics addressed to older adults) has also pinpointed the following recommendations to consider in pre-, during and post-experience: Associate historical information to points-of-interest; share resources, status, and social activities; interlink past/present photographs and other content to the visited places and routes; run pre-trip campaigns relative to the destination and e-guidance for the routes with certified information; provide travel planning information; recommend places and show relevant content to different points of interest based on different criteria (e.g., tourist's location, context, profile); and enable automatic check-ins, tour stamps, city keys, sightseeing, and history clues.

In chapter "Assessing Games for Active Ageing," different methods and data collection instruments for assessing the effectiveness and benefits of games in comparison with other digitally mediated interventions were covered. Game assessment has revealed to be essential to gather gamers' motivations, interaction patterns, and impact of the gameplay experience on everydayness and general well-being and quality of life.

In addition, both player-centric and game-centric approaches should be considered when designing games and providing a good game-playing experience (i.e., rewarding gameplay, sense of fairness, dilemmas, impactful storytelling, accessibility, and context-dependent variables).

When addressing these games to age-friendly environments, heuristics concerning game elements (e.g., reward/punishment oriented conditions, level design, rules) seem to be insufficient and usability, accessibility and assistive components (i.e., to adapt/assist and rehabilitate) are also necessary. However, only combining heuristics and user testing may lead to more thorough explanations on end-users' cognition, and motivation to behavioral change. For that, mixed-methods that combine both quantitative and qualitative approaches may help to cover different lenses of a research problem.

In this sense, a survey design has been suggested to be suitable to assess players' perceived experience, feelings, and attitudes towards gameplay. A survey design may include either questionnaires (e.g., consumer video game engagement, perceived-challenge gameplay interaction or play experience) or interviews that may be context-oriented and be relative to players' preferences and motivations, alongside gameplay and interface testing. Considerations in performing such techniques as A/B testing, group-comparison and participant observations combined with eye-tracking and game logs were also highlighted in this book.

An issue that was not addressed was game analytics and the relevancy of these metrics to active ageing and general well-being, being this is part of our further research work.

In general, we hope the findings of this study to be useful for researchers, game developers and policy-makers, who wish to incorporate games and active ageing in their studies and practices.

Index

About the Authors

Ana Isabel Veloso is the Dean of the Department of Communication and Art of the University of Aveiro and member of the DigiMedia Research Center. Throughout the years, Ana Veloso has been supervising many students of master's and doctoral degree levels. She has also been the coordinator of many research projects, such as SEDUCE 2.0, SEDUCE, EYES ON GAMES and IMP.cubed. She did her PhD in Communication Sciences and Technologies at the University of Aveiro. Before her doctoral studies, in 2006, she obtained her MSc in Biomedical Engineering and BSc in Informatics Engineering from the University of Coimbra. Her research interests include video games, gerontechnology, and other applications of information and communication in technological mediated contexts.

Liliana Vale Costa is a researcher at the DigiMedia Research center and teacher equivalent to invited assistant professor at the University of Aveiro. She holds a European PhD in Information and Communication in Digital Platforms at the University of Aveiro and University of Porto (with internship at the Disruptive Media Learning Lab, Coventry University); an MA degree in Multimedia Communication and a BSc in New Technologies of Communication, both at the University of Aveiro. Her research interests are universal design, digital games, virtual communities, three-dimensional environments, ageing studies, learning, human-computer interaction, computer-mediated communication, natural interfaces, eHealth, mobile apps, and digital inclusion.